What's Luck Got to Do with It?

JOSEPH MAZUR

What's to Do

Luck Got with It?

The History, Mathematics, and Psychology behind the Gambler's Illusion

PRINCETON UNIVERSITY PRESS
PRINCETON AND OXFORD

Published by Princeton University Press, 41 William Street,
Princeton, New Jersey 08540
In the United Kingdom: Princeton University Press,
6 Oxford Street, Woodstock, Oxfordshire OX20 1TW

Library of Congress Cataloging-in-Publication Data

Mazur, Joseph.
What's luck got to do with it? : the history, mathematics, and
psychology behind the gambler's illusion / Joseph Mazur.
p. cm.
Includes bibliographical references and index.
ISBN 978-0-691-13890-9 (hardcover : alk. paper)
1. Games of chance (Mathematics) 2. Chance—
Psychological aspects. 3. Gambling—Social aspects.
I. Title.
QA271.M39 2010
519.2'7—dc22
2009045799

British Library Cataloging-in-Publication Data is available
This book has been composed in ITC New Baskerville
Printed on acid-free paper. ∞

press.princeton.edu
Printed in the United States of America

3 5 7 9 10 8 6 4 2

To Jennifer

the Jackpot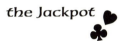

If by saying that a Man has had good Luck, nothing more was meant than that he has been generally a Gainer at play, the Expression might be allowed as very proper in a short way of speaking: But if the Word Good Luck be understood to signify a certain predominant quality, so inherent in a Man, that he must win whenever he Plays, or at least win oftener than lose, it may be denied that there is any such thing in Nature.

—Abraham De Moivre,
The Doctrine of Chances (1716)[1]

You cannot win
if you do not play.
—Proverb

You cannot lose
if you do not play.
—*The Wire*
(Marla Daniels)

Contents

Part III: The Analysis

Introduction

*Everything that happens just happens because
everything in the world just happened.*
—Uncle Herman

When I was a child, my uncles would gather every Saturday at
my grandparents' house to sit at a long dining room table
telling jokes while accounting their week's gambling wins and losses.
My grandfather Morris would retire to a musty back room, crank
up his red-brown mahogany Victrola, lie diagonally across a double
bed with his eyes closed and his feet off to one side, and listen to the
distant voice of the velvety soprano Amelita Galli-Curci sing Gilda's
arias from *Rigoletto*.

Gambling was the family pastime. During those moments when
my grandfather was half asleep on the bed next door, my uncles were
eager to hook me onto horse racing. My uncle Sam would give me a
quarter and betting instructions.

"You gotta analyze," he said one day. "You gotta know the value of
the odds. But the first and most important thing is to be sure that your
horse has a good chance of coming in on your bet. If you don't like a
horse 'cause of its name or 'cause you don't feel lucky that day, don't
bet on it. And don't bet on a horse that's overly backed. Here, look at
Victory Dancer, she's a three-to-one favorite, whaddaya think?"

"I don't like the name," I said.

"Good! Then it doesn't give you good luck. How about—"

I was a very impressionable young boy and took good luck to
be some invisible thing enigmatically passed from one person to

another. There were times when I had it, days when I'd wake with it, moments when I could *feel* it; and then there were times when I sensed it disappointingly gone like the fudge at the bottom of a sundae. One day I won a minor door prize at Radio City Music Hall. What luck! I won a tub of Topps bubblegum and was called to the stage to pick it up—very exciting. That day I had it and felt it with an empowering energy that could have propelled me to the stars. But the next day my bicycle was stolen from a rack of twenty others and I spent the day whining, "Why me?" to myself.

Had it not been for my uncle Herman, I would have believed that the Roman goddess Fortuna turning her wheel impelled the rise and fall of my own luck. On days when I knew my number was on top, I would risk tossing my glass aggie shooter in a game of marbles or buy a few packs of bubblegum baseball cards and feverishly unwrap them, hoping to find a Phil Rizzuto or Mickey Mantle.

Herman walked into the room at the moment Sam was giving me gambling instructions. Somehow my uncle Herman escaped the gambling gene. He was monumentally built, tall, sturdy, and solid with an imposing power of speech and a voice that was both mellow and thunderous, like the roar of a lion that had swallowed a gallon of honey. Herman warned me of the family gambling addiction.

"Watch out for your uncles Sam and Al," Herman said. "They'll give you fifty cents and you're hooked forever. They'll give you advice on a horse that will win and from then on it's in your blood."

"What will be in my blood?" I asked, thinking he meant contamination leading to some foul disease.

"The nasty Hialeah delusion," he said.

"What's that?"

"It's the message whispered in a gambler's ear whenever he comes close to a racetrack."

"What does it whisper?"

"You've got luck on your side," he whispered as if it were some secret meant only for my ears. "Belief that you have luck on your side confuses you to think you have supernatural personal control of what happens to you. But you know there really isn't any such thing as luck to have or not have. Everything that happens just happens because everything in the world just happened."

"They only gave me a quarter," I said, just to set the record straight.

Not completely understanding his little secret, I continued to believe that my own luck gave me absolute control, at least for those days when I had luck, and placed my quarter with Sam to bet on some horse called Brightstar to show. It showed all right. I had no reason for picking one thoroughbred from the next, but there was something in that name Brightstar and something in that day at the races that passed the torch of luck to me. Collecting fifty cents, I felt the heat of the torch and was hooked.

Herman was angry, because he knew that the thrill of that first win was the hook. Now Herman had a funny laugh that seemed to come from his nose—not a snort, but rather a nice laugh that was honest and charming. But when he was angry, his powerful voice would rise in volume and deepen in pitch to assure you that he was irked and that you'd better attempt to change his displeasure.

"Okay, okay," said Sam after Herman boomed out his censure. "Your uncle is right. You were lucky this time. But you should remember that in the end the track wins."

"You should remember more than that," Herman thundered. "Look at your uncles who sit around this radio every Saturday afternoon after they've called their bookies. Do they look rich to you? If they know so much about horses and gambling, why aren't they rich? Don't gamble on a horse that you know nothing about. Even the best horse can slip in the mud and break a leg right in the middle of a race. Even the best horse can be unlucky some days."

As soon as Herman left, gambling conversation would resume. And over my teenage years my uncles would give me lessons on luck and the mathematics of gambling odds.

"Suppose I make a $2 bet on a number that has a hundred-to-one odds of coming up," my uncle Carl once asked me. "And suppose the winning payoff is a hundred dollars. How many bets should I make to give me an even chance of winning?"

"Looks to me that you'd have to bet $2 a hundred times to be sure to win," I answered. "So it would cost you two hundred dollars to win one hundred dollars, not a good return."

"No, no, no! That's not what I'm asking. I'm asking how many bets must I make just to have an even chance."

"Then you must make half a hundred different bets at $2."

"I'd be spending a hundred to make a hundred, no?" Carl said with a laugh.

"Sure," I said. "But there's no guarantee that you'll win after betting on fifty different numbers. You might lose. The only guarantee is if you bet on every number."

"Yes, yes, but I don't want to do that because I'd be losing a hundred dollars. The whole point of gambling is to beat the odds, not match them."

"Well, then, I suppose you just have to bet $2 less than fifty times and take your chances."

"Yes!" Carl said with a contorted smile. "That's called gambling! Ya have to take your chances to beat the odds. Ya have to have luck to win."

The idea for this book came to me late one cold snowy night when my car had broken down just outside a 7-Eleven. Waiting inside the convenience store for AAA to arrive, just before midnight, I watched person after person put down twenty, fifty, even a hundred dollars for Tri-State Megabucks. The pot was up to $2.4 million. From what I could guess, those people were not exactly so well-off that they could blow a fifty on fleeting moments of entertainment—there were some who had me thinking about whether or not their children, if they had any, were getting enough to eat.

For years, I had been demonstrating to my family and students the futility of sweepstakes and lotteries. Then one day it happened. I won first prize, a $20,000 bond, in a Pillsbury Company promotional sweepstakes; my wife had entered my name at a grand opening of a local supermarket. It was a 1.4 million-to-one chance. For one brief interval after that unlikely, fortunate event, I, too, fell under the spell of sweepstake optimism. In the weakness of euphoria, I filled out cards at every opportunity before realizing that my chances of a second win were just as miniscule as the first.

That night at the 7-Eleven awakened an old desire to write about the follies of ambitious belief in windfalls, of lottery optimism, and of gambling folly. A thunderous old pickup pulled up to the front door and a very thin man stepped out of its rusting cab. He entered the store, slapped down five twenty-dollar bills on the counter along with his list of hopeful Megabucks numbers, and said just one word to the young girl behind the counter—"Payday." It was that man and his one word that drove my next thought, which was to write a didactic book on the pitfalls of misunderstanding luck. Later that night, driving home on icy roads, the heater on full blast, I toyed with the idea that the book should be sermonizing the consequences of addictive behavior.

But the book does not sermonize. In rational moments of research and the eventual composition, I found my writing compelled by the history with strong obligations to understand the nubs and cores of gambling, without giving any advice other than that that comes from understanding the mathematics of likelihood as well as an awareness of the risks of gambling optimism. My car breakdown may have been the instrument of the idea, but it would not have motivated me enough without my never-ending desire to comprehend something I had never really understood to my own satisfaction—the magnificent law of large numbers, that theorem describing the long-term stability of the average value of a random variable. It had always seemed too implausible, too fanciful. How could a large collection of independent random happenings possibly be mathematically predicted? How could crowds choose more wisely than small groups or individuals? How could casino managers know their profits in advance with such uncanny precision? How is it that bookies—so sure of the odds they offer—rarely gamble against the odds? And then, weirdest of all: how is it possible that mathematics can be so clever as to compose such a phenomenological law, a principle of chance? These are the driving questions of this book.

Gambling goes back to the beginning of time when cavemen rolled stones the way children in the last century tossed marbles. Homer tells us that gambling and luck have their roots in the beginning of time when Zeus, Poseidon, and Hades drew lots for shares in the universe. Poseidon's luck was to draw the gray sea to live in; Hades got the mists and the darkness; and Zeus the wide sky, the clouds, and light air.[1]

Throughout history, luck has been believed to be not only some desirable thing to be possessed in order to win at gambling but also some telekinetic spirit protecting the body against misfortune. Ancient religions attributed luck to deities of destiny—Min in Egypt, Tyche in Greece, Fortuna in Rome. Medieval tradition personified Fortune as a woman turning a wheel that would determine an individual's fate. An illumination in a fifteenth-century French translation of Boccaccio's *De Casibus Virorum Illustrium*, a collection of moral stories about the rise and fall of famous men, shows Boccaccio pointing to the goddess Fortune turning a wheel of people rising and falling. And Chaucer's Monk's Tale sees luck in Fortune's will:

> For sure it is, if Fortune wills to flee,
> No man may stay her course or keep his hold;
> Let no one trust a blind prosperity.
> Be warned by these examples, true and old.[2]

Such views carry the notion that whatever happens is beyond a person's control. Christianity and Islam designate God as the keeper of luck; his providence may influence a person's destiny. There was a time when these religions accepted omens and ritual sacrifice in attempts to gain favor and influence God's plans; we still use the term *by the grace of God* to mean that God has the power to change destiny and with a bit of luck a person's prayer might get him to do so.

Even today, gamblers are under the impression that there is such a thing called luck and that it can be felt and had with a rabbit's foot or crossed fingers, or lost by the path of a black cat. Maybe they are right, but with the popularity of gambling on the rise and with one in twenty gamblers falling into the pathological end of the spectrum, with reality gambling TV shows such as *Deal or No Deal* captivating huge numbers of viewers, and with Internet gambling and state lotteries sucking up the paychecks of some of the poorest Americans, it seems to be the right time to learn what luck really is, why we think we can possess it with empowering energy, and why we are so influenced by it.

And yet, belief in luck can be a good thing. Its placebo effect is responsible for a good many cures of ailments that would have otherwise overpowered the ill. Not to undermine the helpful feelings of luck and their spiritual benefits, this book concentrates on the mathematics behind gambling to empower the reader who knew— all along—that the powerful illusion of luck is not some acquired supernatural essence but something that can be cogently explained by rules of probability. It alerts and coaches the educated person on the street that that mysterious thing, which agreeably whispers wishful thinking in ears at gambling casinos everywhere, is actually a diabolical con. Ultimately, we want to understand greed and luck in gambling, as well as why people accept bets with negative expectation, and finally answer the fundamental question from both mathematical and psychological positions: what makes us feel lucky in gambling?

There are, essentially, just eight mathematical terms referred to in this book.

1. Binomial frequency curve
2. Combinations
3. Expected value (sometimes called expectation)
4. Mean
5. Odds
6. Probability
7. Standard deviation
8. Standard normal curve

When you come across any of these terms unfamiliar to you, refer to appendix D. Definitions and examples are listed there.

You will notice balloon callouts with numbers in the margins. They indicate that further explanations can be found in appendix E.

Part I

THE HISTORY

Ω

Mathematical models of nature's wonders generally come along after a great deal of experience and observation. Mathematical enlightenment of the wonders of chance and luck arose out of the gaming rooms and coffeehouses of seventeenth- and eighteenth-century Europe, when the common person on the street believed his or her destiny was preordained by astrology, tarot and palm reading were the custom, amulets and crosses were worn to ward off the evil eye, and the momentous ideas of mathematical probability were still in their infancy.

In America the gambling experience spread from New Orleans, where life itself was a game of chance; it moved up along the Mississippi on riverboats paddling from Vicksburg to Cleveland. By 2008, gambling had reached Wall Street, where the hard-won mathematical insights into the complexities of risk behavior were ignored, as even shrewd investors turned into reckless gamblers and financial markets reverted to the naive and reckless illusion of luck and unchecked opportunity. In this modern era, with its hard-won insights into the mathematics of probability and risk assessment, this illusion of luck is inexcusable.

♣ CHAPTER 1 ♣

Pits, Pebbles, and Bones

Rolling to Discover Fate

Gaming is a principle inherent in human nature.
—*Edmund Burke, British House of Commons Record,*
February 11, 1780

Imagine life in the last Ice Age. Those Neanderthals, with their orangutan jaws and beetle brows, burbling some mono-vowel language, sharpening spears in preparation for a hunt of hungry scimitar-toothed black tigers, reflexively gambling every day against the impending extinction of their race.[1] Ground tremors, as common as cloudy days, triggered by great weights of melting ice continually relaxing gargantuan pressures of the earth's indomitable crust; ordeals of menacing elements, snow and freezing temperatures; hunger, pain, and weakness from the bruises of long, fierce hunts; and most worrisome of all, the daily threats of nearby ravenous beasts stealthily looking for supper. Humans have been gambling ever since that unfathomable island of time, when herds of pachyderms and hump-shouldered mammoths freely wandered over the frozen lakes of the Neander Valley.

Our extinct fellow proto-humans looked brutally intimidating and menacing with their powerful muscles, fleshy fingers, and massive limbs, but they were innocently carrying fierce looks for passive self-protection.[2] They cared for their young, who played with discarded dry bones by kicking, banging, and tossing them here and there in amusement or quite possibly in a shilly-shally appreciation of primordial sport. How natural it is for a child to create games by hurling things. We can readily imagine a Neanderthal child uttering, "My bone for your two that I can throw mine further," even in a primitive language of consonants mixed with *ah* as its only vowel.[3]

We can picture the adults, in their rare, brief leisure time, wagering on who can throw the farthest spear or on who can down the nearest rhino. They may have tossed bones in games the way kids now toss marbles, laugh when something is funny, or cry when injured. Neanderthals smiled to express joy, frowned to convey displeasure, embraced in camaraderie, and gambled every day with their own lives in decisions of whether or not to go out on a hunt or to wander far from their recognized comfortable safety zones where the fire was warm.

Risks are the gambles, the games, the balance of expectation and fate. And luck rarely comes without risking the possibility of loss, injury, trouble, vulnerability, ruin, or damage in a universe of opposing chances. We also know (from bone sample evidence) that our Neanderthal friends were subjected to a high rate of injury during their lifetimes, most likely from close-range hunting of fast and ferocious sabertooths, whose sword-like canines could effortlessly pierce and slash the skin of even the toughest males. And if the cats—those felidae with the courage to raze mighty mammoths—didn't occasionally slash those hardy men to death, then the bruises those cats inflicted surely disabled them. That was the true gamble—to eat or be eaten.

Gambling is about odds, the chances of things going one way or another. Will the team bring down the mammoth, or will a tusk lethally impale someone? So we humans are programmed to gamble. It's not only about the team. We take risks, leave our houses and explore the uncertain boundaries of secure and reliable neighborhoods, all as part of the animal nature of survival. The urge to take risks is just one of the hardwired universals of being human, along with smiles, frowns, cries, and laughs.

When we come closer to our own time—not much closer, still in the late Pleistocene era, 10,000 to 40,000 years ago—we find our early dark-skinned Cro-Magnon ancestors in the Danube Valley painting on the walls of caves with sticks of charcoal, carving moon calendars or in performance of some ritual event to applaud the supernatural owner of the wildlife they hunt or to thank the shamans.[4] They, too, performed risks and had to ask: *Is it safe to paint today? Should we leave home and take care not to be mauled by the nearest beast? Or should we stay protected by warm fire and eat the spoils of yesterday's kill?* It was all a gamble at a time when humans were skillfully tuning maneuvers of feet, arms, and wrists to influence and direct flights of their sharp weapons.

Their tools, those spears, arrows, flints, and fires, gave them hunting advantages their brawny Neanderthal neighbors never had: the opportunity to hunt from safe distances. And with them came prophesizing games and innocent gambling. Innocent, because players were not necessarily wagering their fine spears or furs nor—what should have been quite reasonable—staking their best pickings, but rather banking on the moods of randomness for providential guidance and help from the phantoms of predictability in forecasting decisions. A shaman might roll a pair of, say, sheep astragals (anklebones) to determine if the tribe should go out on a hunt the next day. Die-like objects such as filed and sanded astragals have been found in abundance at archaeological sites almost everywhere from central France to as far east as the Punjab. What they were used for is anyone's guess, but more likely than not they brought some form of entertainment or a means of communication with the gods.[5] Someone would ask a question, and, depending on how the astragals fell from the shaman's hand, there would be an answer. One answer was accepted when wide sides faced upward, another when the narrow. Surely these bones were biased; however, it did not take long for our clever ancestors to find a way of evening the odds by rasping the sides of an astragal and smoothening out six faces of a stone or piece of wood for fairer outcomes in inventing the die. Certainly, sticks and odd-shaped stones must have been used for playing against chance. Fruit pits, pebbles, shells, teeth, seeds, and acorns must have given hints of rolling to discover fate.

When we come to the more modern hunting and gathering societies of 10,000 years ago, we see that gambling advances with better impressions of randomness. We find dice being made from carefully carved pieces of wood, stone, and ivory. It's more than just the archaeological findings; now we have the literature, albeit through oral tradition, to back it up. Homer's *Iliad* tells us that gambling and chance have their roots in the beginning of time when Zeus, Poseidon, and Hades drew lots for shares in the universe.

Lot is the etymological root in the words *allotment* and *lottery*. It is also something that happens to a person when the *lot* has fallen—*it was his lot*. The *casting of lots* would have meant any decision-making procedure or mechanism, the flip of a coin, the roll of a die, the pick of a straw. The *lot* itself might have been an object such as a piece of wood, a pebble, a die, a coin, or a straw that could be used as one of the counters in determining answers to vital questions by the position in which it comes to rest after being tossed or picked. But if the lot is to be fair, it must be far more unbiased than an astragal, which surely does not have equal chances of falling on any of its four sides.

Drawing lots was thought to be the fair way to settle a choice that could not be established by reason. And since every lot banks on the whim of Fortune, or on the very misunderstood impulse of randomness, it might be said that every unreasoned choice is a gamble. Indeed, children of all ancient cultures must have muttered some variant of *eenie meenie mynie mo* among friends making a choice. Still, for the adults, it was more likely thought of as a means of communication with some supernatural spirit. Getting the short end of the stick might have been the random pick of the draw, but it could also have simply been the will of God. The Mishnah (the section of the Talmud connected to oral laws) says that *to draw lots* one must have an urn of tablets marked to describe a fate and that those tablets must be alike in size and shape so that any pick is as likely as any other. The Bible says that to atone for the sins of his house, Aaron was to cast lots to decide which of two goats would be sacrificed and which would be sent back into the wilderness.

In Exodus there is a vague description of *the breastplate of judgment,* a part of Aaron's priestly garment. Aaron was to wear it on entering a holy place to seek God's judgment on difficult questions affecting the welfare of Israel.[6] The breastplate of judgment refers to a vestment of embroidered fabric set with twelve precious stones representing the twelve tribes of Israel and worn by any of the high priests seeking God's guidance on matters concerning the welfare of the community. According to instructions outlined in Exodus, the breastplate must contain the so-called Urim and Thummim, which literally translate as "the Lights and the Perfections." According to some modern commentators, these were two sacred lots, chance instruments such as coins or dice, used for the purpose of determining the will of God on questions of national importance.[7] It may have been that the priest would cast the lots but also understood that while the lot is cast, God manipulates the lot to determine the outcome—"The lot is cast into the lap; But the whole disposing thereof is of the Lord" (Prov. 16:33).

We take "the Lights and the Perfections" here to mean the perfect determination of the truth by means of unbiased casting of lots—the perfect throw of perfect dice for the fairest possible decision.

Fairness is, unfortunately, seldom a functioning human trait, but when it comes to decision-making, inequality is inherently recognizable. The child who must share a piece of cake after being given the opportunity of dividing and cutting it to permit others to choose pieces will try to divide and cut with fairest precision. Humans can recognize overt inequities. So it shouldn't have taken many rolls of astragals to upset early gamblers and cause them to think of a fairer, more random way to cast lots. Though the typical astragal is shaped somewhat like an elongated cube, it has only four sides to fall on; its two end faces are so uneven and knuckly it would be highly unlikely for it to remain standing on one face. Yet, astragals were used for centuries before real cubical dice took over, sometime long before the first millennium BCE, when cubes that could (more or less) fall fairly on one of six faces were introduced. Since then, dice variations have been used in every part of the world from America to Japan, from Sweden to Africa. Recent (2004) archaeological digs in the Bronze Age city of Shahr-I Shokhta (literally, the Burnt City) in southeast Iran unearthed a five-thousand-year-old backgammon set made of ebony with cubical dice.

Dice playing enters stories such as the ancient Indian Sanskrit epic poem *Mahabharata*, the fifth century BCE tale of Cyrus the Great, and the story of Isis and Osiris in which we learn of the Egyptian game of *tau* (akin to English draughts or checkers, which goes back to at least 1600 BCE). Older still is the Royal Game of Ur, going back to before 2500 BCE.[8] Two players would race their pieces from one end of the board to the other according to moves controlled by specific landings of a die, a prototype of backgammon. The die would have been either a four-sided stick or tetrahedron.

Modern dice, numbered as ours with opposite sides summing to 7, have been found at archaeological sites in Thebes and elsewhere in Egypt.[9] And we have evidence that the Egyptians played a game called *atep*—a game still played almost everywhere in the world and one I recall playing as a kid when we had to either choose sides or choose who would go first in a game. We'd call out *odds* or *evens* and then extend either one or two fingers on the split second after calling out *1-2-3-shoot*.[10] For such a game there are no physical lots but rather mental choices (as if flipping two coins at once) to make fair decisions.

When we come to the Romans we find gambling rampant, though we also find evidence for the first laws against gambling to dampen uncontrolled behavior associated with gaming. It was a time steeped in sexual marathons and drinking sessions mixed with gambling by dice, cards, and quail fights. In nineteenth-century archaeological excavations of Rome, hundreds of gaming tables were found. The tables were typically designed as *tabula lusoria* (table of play) in the form of three horizontal lines, each containing twelve signs with words arranged to make a sentence with thirty-six letters. The taverns patronized by gamblers used such poetic forms in their signs to warn against fighting over games, no doubt swayed by the thirty-six (6 × 6) distinct possible results of throwing two dice.

LEVATE	LVDERE
NESCIS	DALVSO
RILOCV	RECEDE

These six terms with thirty-six letters are abbreviations of words that form a sort of haiku rune that unravels to this rough translation: *Rise! If you don't know the game make room for better players*!

FIGURE 1.1. Royal Game of Ur, southern Iraq, ca. 2600–2400 BC. From the British Museum Room 56: Mesopotamia. Copyright © The Trustees of the British Museum.

Such *tabula lusoria* would signify good or bad luck, warn of the skill needed to play well, or warn of the risks of gambling. Others were unguarded invitations to gamble.

We know from Plutarch that Marc Anthony was a consummate gambler; that Augustus was an ardent dice player; and that Nero played some variant of craps. Claudius had a special carriage designed for playing dice; he even wrote a book on dice playing. And Caligula, after losing all his money at an ancient variant of craps, ran into the street, confiscated money from two Roman guards, and returned to his game.[11]

Dice have been found all along trails used by the crusaders. Throughout the Middle Ages, from northern Europe to Brindisi at the heel of Italy, crusading armies played dice games at taverns along their way.

Late in the thirteenth century, Alfonso X, king of León and Castile, commissioned the writing of a book of games. Known as *Libro de*

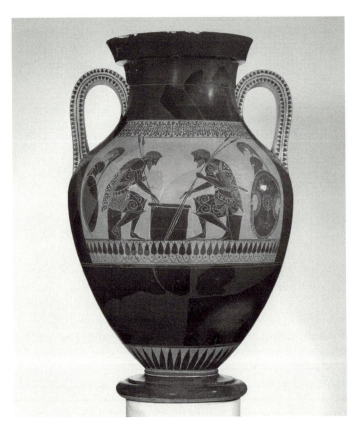

Figure 1.2. Achilles and Ajax playing dice (Attic bilingual amphora), ca. 525–520 BC. Reproduced courtesy of the Museum of Fine Arts, Boston.

Los Juegos (Book of games), it contains descriptions and illuminated illustrations of all sorts of games from chess to backgammon, including dice and tables.[12]

The story told in Alfonso's book of games is that there was once a king who would often consult his three wise men over the nature of things, and on one particular occasion the debate came to the question of gaming and of the advantages of luck and brains.[13] One wise man said that brains were more valuable than luck because thinking gives order to life and even if he lost, he would not be to blame because he used reason. Another said that luck was more valuable

FIGURE 1.3. Image from folio 75 verso of the *Libro de Los Juegos* by Alfonso X, depicting two ladies playing the game of *seis, dos, y* (six, deuce, and ace). Each player started with fifteen pieces. In this illumination one player has 8 on the sixth column, 4 on the second, and 3 on the ace point. The other player has 5 on each of the remaining three columns. The players are moving their men in one direction around the board (as in backgammon) to get to the opposite side of their starting positions. Note that three dice are being used. Reproduced courtesy of Bridgeman Art Library.

than brains because win or lose, his brains could not avoid his destiny. And the third said it would be better to have both—to use the brain to his reasoning advantage and to use luck to protect him from any potential harm.

Alfonso was on to something that was to become the core understanding of all professional gambling from cards to hedge funds. The balance of luck and reasoning could be interpreted through a rational measure of how favorable the outcome might be. Though Alfonso had no conception of risk management and certainly no perception of positive expectancy (the mathematical tool that modern professionals use to quantify a future event), he did understand that the blends of luck and skill behind gambling games fall into a wide spectrum.

Alfonso's book of games tells us that dice are perfect cubes, made of wood, stone, bone, or, best of all, metal—as perfect as thirteenth-century craftsmen could manufacture—otherwise they would roll more often on one side than any other and would be a trick of luck. The spots are marked just as they would be on our modern dice with opposite sides summing to 7, but for some reason—possibly to acknowledge the holy trinity in hopes that they may have some influence—games were played with three dice.[14] The games were simple: in *mayores*, he who rolled highest won; in other games, he who rolled lowest won.

Alfonso's game book makes the point of saying that many games of the day were designed to resemble events and customs of the times, showing how kings during war would fight alongside their soldiers or how individual soldiers of other kings would be killed, captured, or expelled from the land.

> And also as in the time of peace they are to show their treasures and their riches and the noble and strange things that they have. And according to this they made games. Some with twelve squares (per side), others with ten, others with eight, others with six and others with four. And thus they continued descending down to just one square, which they divided into eight parts. And all this they did because of the great similarities according to the ancient knowledge, which the wise men used.[15]

There was still no notion of a mathematical measure of likelihood. Such an exotic concept would have required knowing something about permutations and combinations of objects, a subject that was almost entirely unknown. A permutation of n objects means all of the possible distinct arrangements of those n objects; so, for example, *ABC* is considered different than *ACB*. A combination of n objects means only that n objects have been selected; therefore, *ABC* is no different from *ACB*. These are two critical notions that are at the heart of gambling, for every random event entails several possible outcomes with or without regard to order.

FIGURE 1.4. The eight combinations of two symbols.

The Chinese were already engaged in some thoughts on permutations and combinations. The *I-Ching* (Book of Changes) entertains a symbol system to identify order in random events. It goes back to the third millennium BCE and may be the earliest work dealing with *combinatorics*, the branch of mathematics dealing with combinations and permutations of members of a group of objects and the mathematical relations that characterize their properties. We take the yin yang symbols to represent 0 and 1. If the solid bar (yang) represents 1 and the split bar (yin) represents 0, then these so-called trigrams are simply binary representations of the numbers 0,1,2,3, . . . Moreover, from just combinations of two symbols (the solid line and the broken line) taken three at a time, we get eight distinct objects.

Abstractly, these eight objects represent all the combinations that can be made by taking two things in groups of three. As an example of how that may play out in a gambling game, take the game of flipping three coins (heads yin, tails yang) and wagering exactly two heads will appear. The probability of getting exactly two heads is 3/8, since there are three groups with exactly two split bars.

As for the Greeks, aside from the few cases of combination counting that we learn through Plutarch, it seems that they never developed a systematic theory of combinatorics. Plutarch tells us that in the fourth century BCE, the Greek philosopher and mathematician Xenocrates computed the number of different combinations of syllables in sensible words of the Greek language as 1,002,000,000,000.[16] (Xenocrates' calculation must have been based on finding the number of possible syllables in the Greek language, surely a daunting lexicographical as well as mathematical exercise.) In the sixth century, however, the philosopher Boethius (who was recently elevated to sainthood by Pope Benedict XVI) figured out that the number of combinations of *n* things taken 2 at a time is simply

$$\frac{n(n-1)}{2}.\text{[17]}$$

This is a very easy calculation, if one looks at it as simply writing out the n things twice. Write out the two strings of n objects as

$$1, 2, 3, \ldots n$$
$$1, 2, 3, \ldots n.$$

Then notice that to pair two at a time would mean pairing each number of the first string with all the numbers of the second. That gives you $n(n-1)$ parings. Note that the $n-1$ occurs because you have to eliminate n pairings of a number with itself. But notice that you have counted twice, and so you must divide the result by 2 to get

$$\frac{n(n-1)}{2}.$$

One might say some of the essential mathematics of gambling were around as far back as the eighth century with the interest in Jewish mystical writings that calculated various ways in which the letters of the Hebrew alphabet can be arranged and came up with the correct colossal number.[18] The twelfth-century Spanish biblical commentator Rabbi Abraham be Meir ibn Ezra carried out some of the earliest of the impressive calculations of combinations. Ibn Ezra was able to calculate the number of combinations of 7 objects taken k at a time where k could be anything less than 7. His interest was the possible conjunctions of the seven known planets, which then included the sun and moon. The twelfth-century Indian mathematician Bhaskara extended Boethius's computations in his arithmetic textbook *Lilavati* (The beautiful), written for his child, Lilavati, giving rules for finding the number of ways to choose a group of r objects out of a group of n objects, and posing questions in illustrative mathematical narrative.[19] And surely, the fourteenth-century work of the mathematician and biblical commentator Levi ben Gerson, *Maasei Hoshev* (Art of calculation), should have been a significant contributor, as it correctly demonstrated the general formula for the number of combinations of n things taken r at a time, the principal tool in calculating odds.

However, this was a time when great ideas emerged in isolation, from biblical commentators who worked within the confines of a village, monks who never left their monasteries, and mathematicians who rarely circulated more than a single copy of their work. And so, alas, these contributions were unknown to the collaborative scientific communities of Europe and had to wait centuries before being rediscovered as if new.

They were rediscovered in the mid-thirteenth century. A manuscript of the Latin poem *De Vetula* was found in Ovid's tomb and immediately became a medieval best-seller. It was copied and distributed to libraries all across Europe, though very likely not written by Ovid himself. The poem itself is autobiographical, three books about a poet who changes his lifestyle because of a regrettable love affair. Leading a licentious life (described in detail), he has an affair and falls in love with a beautiful woman. When her husband dies twenty years later, he marries the woman and discovers that she is now old and that he was conned. Depressed, he turns his life to more lofty and moral pursuits of mathematics, philosophy, music, and, of course, religion. In the first book he gives a discourse on the laws of chance applied to gambling with three dice along with his reasons for avoiding dice games.[20]

Though the poem is a medieval morality verse, it does give evidence that some basic mathematical rules of permutations and combinations were known at the time of the discovery of the manuscript, at least as far back as early fourteenth-century France and quite possibly much earlier in India, since the knowledge likely came from Arabic and Indian sources. Regardless of its authenticity, the *De Vetula* contains the earliest known calculations involving serious probability through the observation that in the random throw of dice certain numbers have more ways of occurring than others—the smallest and greatest sums occur more rarely than those near the mode, the most frequent value, just as they do for a pair of dice. (See figure 2.1.)

By Henry VIII's time gambling was illegal in England and there were ordinances against gambling in many European countries, but at court almost everything seemed to be legal. Kings and queens could play as they wished as long as it was in the private apartments of royal residencies. It was customary to announce "His Majesty is

FIGURE 1.5. From an edition of Boccaccio's *De Casibus Virorum Illustrium* (Paris, 1467), MSS Hunter 371–72 (V.1.8–9), volume 1, folio 1r. Lady Fortune with the Wheel of Fortune. As the wheel turns some men may rise from poverty and hunger to greatness, while some great men may fall. Such scenes of the rise and fall of man were typical in the Middle Ages.

out" when the king entered a game; with that announcement, it was understood that court formality, ceremony, and etiquette could cease. When it was his pleasure to discontinue the game, it was announced "His Majesty is at home," whence playing would cease and the ceremony of the palace was returned to normal.

A parliamentary act passed under Henry VII forbade gambling at any time of year except during the twelve days of Christmas. During those twelve days the public was not only permitted to gamble but encouraged to do so in church.

> Whereby they thinke, throughout the yeare
> > to have good luck in play,
> And not to lose: then straight at game
> > till day-light they do strive,
> To make some pleasant proofe how well
> > their hallowed pence will thrive.
> Three Masses every priest doth sing
> > upon that solemne day,
> With offerings unto every one,
> > that so the more may play.[21]

Like the Romans, Elizabethans were eager gamblers. Despite legal obstacles that continued up to the time of Elizabeth, we know from Shakespeare's plays that gambling popularity was widespread before and during the Renaissance. Gamblers flocked to the vibrant city of London, where festivities lasted through the year, a city where the individual could lose identity and escape into rhythms of fantasy.[22] Society and royalty made no attempt to conceal gaming. Sir Francis Drake, Thomas Digges, William Gilbert, and Ben Johnson frequently gambled at hazard (the popular seventeenth- and eighteenth-century forerunner of the dice game craps)—it was, after all, the social norm of gentlemen. Christopher Marlowe, Thomas Middleton, Sir Walter Raleigh, and the queen herself often played tables and hazard, and occasionally wagered in the popular blood sport of cockfighting.[23] Shakespeare saw gambling as an integral part of his world and—like all his other apt observations of human eccentricities—he skillfully used it for suitable metaphors.

Fate was directly linked to the mechanical movements in the sky. In *King Lear*, the bastard, Edmund, struggles with the connection between his bastardhood and movements in the sky. Even as late as 1606, when Shakespeare wrote his play about the mythical impetuous old king, the sky was thought of as a fixed (firm) canopy, studded with diamond-like stars, an impression that lingered through the centuries ever since Aristotle declared that the stars influence even birth and extinction.[24] People still believed in a mechanistic, deterministic universe where the notions of destiny, action, and reaction were indubitably linked. And what about *The Tragedy of Hamlet*? Elizabethans, and even Jacobins, would have no trouble believing that the murder of Hamlet's father was the cause of all that followed— Hamlet's madness, Ophelia's drowning, and, ultimately, the deaths of Gertrude, Laertes, and Hamlet—but the murder itself, *that* would have been initiated by the movements of the diamond-studded crystalline spheres nested, one in the next, with the earth at the center, each with a glowing jewel set in its transparency, all moving in perfect musical harmony. It is easy to understand how the impressions of fate seemed mechanical to a person who believes in a finite universe. It is hard to imagine the thoughts of an Elizabethan lying face up in a country field on a dry, warm, and moonless night, staring at the vast and wonderful Milky Way, yet he or she must have believed the universe finite and wondered about its size and how the mechanism of its laboring motion determined his or her luck.

♣ CHAPTER 2 ♣

The Professionals

Luck Becomes Measurable

On one nice Trick depends the gen'ral Fate,
An Ace of Hearts steps forth; the King unseen,
Lurk'd in her Hand, and mourn'd his captive Queen.
He springs to Vengeance with an eager Pace,
And falls like Thunder on the prostrate Ace.
—*Alexander Pope, Rape of the Lock*

In rational moods we envision fate as a random, aimless, unmanageable power that causes luck to turn positively or negatively in the interest of an individual or a group. We may envision Fortune turning her wheel, making use of nature's tools of uncertainty to determine the result, or some divine spirit tossing a coin. But, too frequently, our specious present stretches that vision to more supernatural meanings that at times embrace demons and at other times angels. We are *in luck, out of luck,* or *down on our luck.* Whatever it is, we think of it as material, something to capture, be in, be out of, or possess! It rarely comes without risk. And risk implies the possibility of loss, injury, disadvantage, or destruction, something that creates or suggests a hazard or adverse chance.

The Jews associated luck with magical powers of the number 7; the number of days of the week, determined by God's Sabbath, may have been arbitrary, but what about the number of moving objects in the solar system visible to the naked eye or the colors of the rainbow? What about the only possible number markings on the faces of that perfect solid, the cube, where the sum of the numbers on opposite faces is always the same? Historians have been quick to point out that the Greeks and Romans lacked an understanding of chance. It may seem to us surprising that all during those past centuries of gambling there was no mathematical theory of gambling odds. Greeks and Romans simply cast their dice, believing that either the luck of fortune or some god determined their fate (assuming that the *De Vetula* was in fact written many centuries after Ovid's death). Though their understanding of a sophisticated likelihood of casting a particular number may have been wanting, surely they knew that certain numbers were more likely than others, surely they knew that 7 comes up more often than any other number. All they had to do was count the number of ways a 7 could come up against the number of ways any of the others could. There are 6 possible pairs that sum up to 7 and only 1 way that a 2 can appear. (See figure 2.1.)

Such an accounting is at the heart of probability theory. Even today, the appeal of dice is in its mystery. Knowing the chances of dice falling on any particular number does not diminish the thrill of the outcome. However, before the beginning of the sixteenth century, mathematics was coupled with certainty and reserved by natural philosophers and mathematicians for the serious things in life, either to understand the abstract notions of number theory and geometry or applied to explain the practical and functional, such as surveying and other building practices (especially for cathedrals). The mathematical enlightenment of chance was on the verge of emerging by the sixteenth century in the unpublished papers of Girolamo Cardano titled *Liber de Ludo Aleae* (The book on games of chance), which turned out to be recognized as containing the essential elements in understanding the nature of chance and modern probability.[1]

Girolamo Cardano was a Milanese physician, mathematician, and gambler, better known for his 1545 published book, *Ars Magna* (The great art), his account of everything known about the theory of algebraic equations up to the time.

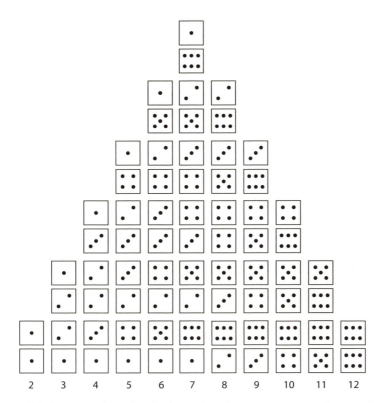

FIGURE 2.1. The number of pairs in each column represents the number of ways each number can appear.

It seems evident that Cardano had no direct intention of publishing his *Liber de Ludo Aleae,* for it was a rambling of mathematical and philosophical jottings. Yet providence had other plans. The manuscript was short, just fifteen pages, but—like other powerful monographs of history—it positioned in place concepts that are now considered the cornerstones of probability theory: expected value, averages and means, frequency tables, the additive properties of probabilities, and calculations on the combinations of ways of having *r* successes in *n* trials, and even gave an obscure sketch of what would later become known as *the law of large numbers* (which essentially says that the probability that the actual average success ratio differs from

FIGURE 2.2. Girolamo
Cardano, 1501–76.
Copyright © 2009 Photo
Researchers, Inc. All
Rights Reserved.

the mathematically predicted success ratio is as close to zero as we
wish, provided that the number of events can be taken as large as
needed to force that condition).

A more thorough analysis of the law of large numbers will be dealt
with later. For now we'll think of it more vaguely as a principle that
suggests how averages tend to behave in the long run. There are only
two sides to a penny and heads is one of those sides, so the probability of a penny coming up heads when flipped is 1/2. According
to recent experiments, the tossing of a real penny is entirely determined by the physics of the spin and not by any random behavior.
It is, however, influenced by imperceptible human controls and tendencies that make its behavior seem random. In one experiment a
machine flipped a penny in such a way that heads came up on every
try.[2] For a real penny, there is an imperceptible difference between
the head and tail surface pressings, but not enough of a difference
to bias the physics—not so with the mathematical penny. That may

seem to suggest that if tails comes up too often on the first few flips, we should expect heads to come up more often on the next few. But there is that strange law called the law of large numbers, which presumably tells us (though it surely does not) that if you flip a coin long enough we should be confident that half the time it will come up heads and half the time it will come up tails. So suppose that 75 tails and 25 heads are the result of a hundred flips. Does that mean that for the next hundred flips we should see more heads than tails? No! The penny does not keep a running tally of how many tails and heads appeared before. It doesn't even have the short-term memory of what happened on its last flip! This bizarre, yet magnificent law does not say what many amateur gamblers believe when they confuse it with the *law of averages*, that nonsensical law that says half the time the coin will come up heads and half the time tails. Rather, the law of large numbers says that after flipping a large number of times there will be a tendency—*a tendency*—for the number of heads to be close to the number of tails. It does not say anything about how large the number of flips must be, only that as the number of flips grows large, the ratio of the number of heads to the number of tails will probably come closer and closer to 1.[3]

The confusion comes when we talk about odds. In rolling a pair of dice, we say that the odds of rolling snake eyes (two ones) are 35 to 1, meaning that there are 35 unfavorable outcomes to 1 favorable. That may seem to say that if you roll a real pair of dice it will take 36 rolls to get one snake eyes. But not quite. The law says that you are *much more likely* to roll snake eyes one hundred times if you rolled the dice 3,600 times. And, of course, you are even more likely to roll snake eyes a million times if you rolled the dice 36,000,000 times.

It is a hazardous law that gives false impressions responsible for a great many gambling mistakes. By presuming that an event remembers its history we make the infamous gamblers' mistake, called the *Monte Carlo fallacy*. Like my uncle Herman's *Hialeah delusion*—which Herman claimed whispered to everyone at Hialeah *you've got luck on your side*—the Monte Carlo fallacy whispers into the novice gambler's ear at roulette wheels of casinos everywhere—*After a long run of red, bet on black*. The novice gambler is compelled to place bets on black after a long run of reds. Placing your bet on 0 puts your chances of winning at a probability of 1/37, so you are more likely to win by choosing one of only two possible colors rather than one of thirty-seven numbers. Would you bet on 0, if 0 did not come up in the last hundred spins? The probability that it would come up on the hundred-and-first spin is exactly the same as it was for any of the first hundred spins, yet players will look at the board that posts past outcomes to make their decisions as if somehow they are handicapping the odds of a horse race. But you might bet on 0, if you believe that in a fair game 0 should turn up more frequently than once in a hundred spins.

Cardano's manuscript, including his version of the law of large numbers, was very likely responsible for the explosion of gambling practices of the seventeenth century when gambling evolved from an activity between naive players—innocent to chance—to professionals and bankers who knew how to consistently win in the long run by playing expected values to their advantage. Moderately harmless—though often raucous and scandalous—social gaming rooms and taverns turned into enterprising casinos with house advantages and professionals who knew how to play the odds.

Cardano posthumously endowed the world with the seeds of probability theory and answered the ancient question of what—besides Tyche, Fortuna, or some other divine intervention—caused chance to favor one outcome over another. In his little manuscript was the germ of a science of chance: from a given set of observable facts we find out how to get a number that tells us what is likely to happen and, according to the renowned French mathematician and physicist Henri Poincaré (1854–1912), the world learned that a person has the same chance as any other person and even the same chance as the gods.

Of course the theory of probability was bound to emerge sooner or later. Almost half a century after Cardano's death, Galileo wrote a short treatise on the odds of throwing three dice and solved the curious mystery of why 10 and 11 seemed to appear more often than 9 and 12.[4]

By listing the combinations (in a similar manner to figure 2.1) Galileo found that there are 27 distinct ways for three dice to sum to

10 and 27 ways to sum to 11, but that there are only 25 ways for three dice to sum to 9 or 12.[5]

Surely it didn't take Galileo's mathematics to convince gamblers that there was a fundamental understanding of dice outcomes. Centuries of observation and practice produced a folk knowledge of dice. By this time dice gamblers had an instinctual knowledge of the odds. Surely they knew that 10 and 11 turn up more frequently than any other number, but once those old instincts became grounded in mathematical explanations and connected with the confidence of being explained by mathematics, gambling took a new direction and became a business that bankers could count on. For the professionals who knew the mathematical odds, gambling was no longer a risk. In the long run, it was *almost* a certainty.

If you were a sixteenth-century professional gambler, you would have done well with some capital, a table, and three fair dice. You might have given your customer 50-to-1 odds at rolling less than 5. There are just four ways of rolling less than 5.

$$(1,1,1),\ (2,1,1),\ (1,2,1),\ (1,1,2)$$

And there are 216 possible combinations and therefore the odds of an unsuccessful roll are 212 to 4, or 53 to 1. (See appendix D for computing odds from a known probability.) Since you would have known the mathematical odds are 53 to 1—that is, a player will score just once in 54 tries—you would be ahead by 1 unit for every 54 rolls of the dice. You would have to pay out $50 for every dollar in play, but the player would have to roll the dice (on average) 54 times (on average) to win $50. Such a margin doesn't seem like much, but keep in mind that you would expect many players to take you up on what looks like a fabulous odds offer. And if you could, you might have pushed the odds more in your favor by lowering them to, say, 30 to 1, though by doing so you might have discouraged some customers from playing. Adjusting the odds to suit your expectations may be tricky but just a game of mathematical tweaking.

Gambling may go back to the beginning of time when cavemen rolled stones the way children in Joe DiMaggio's time flipped baseball cards. It may have been innocent social entertainment, but by

FIGURE 2.3. Young girls playing with astragals. Detail from painting *Children's Games* (1560) by Peter Brueghel the Elder (c. 1525–1569). Reproduced courtesy of the Künsthistorisches Museum of Vienna.

the mid-seventeenth century it was quickly approaching the top of the charts of pastime entertainment. With all the new possibilities, from cards to hazard, it couldn't help but burgeon into a frenzied stimulant akin to the excitement of a horse race.

The serious question for the gambling artist was how to secure those rewards of chance. Mathematics had done a pretty good job of supporting cathedrals, holding up bridges, and predicting eclipses. Gamblers needed a mathematical theory of chance.

That theory emerged in Paris during the cold winter of 1654 when even the Seine froze. People skated and slid on the river while fires burned at street corners and parish priests distributed bread to the poor. It was unusually cold for Paris. Those who had money suppressed their spending on themselves to provide food and warmth for the poor people who could not find work that difficult winter. Thirty years of religious wars in Europe had drained the treasury of France. With huge debts and expenditures accumulating at alarming annual rates, France was forced to increase taxes on its working classes. Dishonest tax collectors and agents brought very little revenue back to the treasury. Exempt from taxation, the nobility continued to accumulate appalling excesses of wealth. It was the early reign of Louis XIV when France was becoming the leader of Europe, a time when the idle rich were overtly gambling in gaming rooms all over Paris.

The typical gambling room consisted of just one table surrounded by two circles of not particularly comfortable armchairs. Heat came from a glowing coal fireplace and light from both candles and smoky, smelly animal fat lamps along windowless walls. They were dark rooms with sometimes as many as fifty people standing, milling, and enjoying snuff. The more refined establishments had more relaxed surroundings—comfortable chairs upholstered in silk, satin, velvet, or damask; carved ebony leather-topped gaming tables with ivory inlay and gilded trim, their multiple inlaid playing surfaces interchangeable for backgammon, checkers, chess, cards, and hazard; and less smelly, less smoky sperm oil lamps. The walls were generally made from carved mahogany panels with extravagantly ornate, florid, and convoluted trims, and though they would be lined with many gilt-metal ornamental candle sconces, the rooms were just as dark as those of any other gambling establishment. The men would relish their snuff and brandy. As for women, they frequented the more aristocratic establishments—a few widows venturing small sums and others, who were under no obedience to a husband or father, quietly observed, flirted, and seduced.

In the mid-seventeenth century the population of London was somewhat smaller than that of Paris. By the end of Queen Elizabeth's forty-five-year reign, England had expanded into a booming economy with a professional class of bankers, lawyers, writers, and scientists. Taverns, alehouses, and sitting rooms and gaming rooms were smoke-filled. Entertainment was everywhere. The gambling houses were quiet compared to the crowded, narrow, noisy streets teeming with vendors, thieves, pimps, whores, and beggars. London was one of the few places in England where a person could be anonymous. In consequence the gaming houses of London were places not only of escape but also of fantasy. London was the perfect place for entertaining gambling, and gambling houses opened by the score all over the town.

Many of these establishments made their money by selling wine and snuff, and took a cut of the stakes. They employed croupiers—sometimes four to a table to spot *punters* (amateurs)—waiters, orderlies that scouted for approaching police, *flashers* that spread false rumors about how much the bank was losing, and *puffs* or *Shylocks* who loaned seed money to hook unsuspecting novices. The larger establishments had armies of paid employees, including ruffian *dunners* who went after those who welshed. The men and women who frequented these establishments were not professionals, for there was no seriously viable theory of gaming available for studying how to become one, apart from a few time-honored intuitive rules and simple counting principles of mathematics. By professionals, I mean honest professionals. Without the mathematics of odds, the only easy way to be sure to win was to cheat, and cheating was rampant. But this was about to change.

On that particularly cold winter afternoon of 1654, the distinguished nobleman and notorious gambler Antoine Gombauld, known by his nobility title as Chevalier de Méré, was on his way to court when he ran into his friend, the amateur mathematician Pierre de Carcavi. De Méré knew an old gambling hunch, not mathematically supported, that suggested favorable odds on betting that he could throw at least one double six with twenty-four throws of a pair of dice. His fortune was quite considerable but he was beginning to lose heavily at the gambling tables and was wondering why the old gambling rules were no longer working in his favor. He confessed his growing gambling addiction and the huge toll it had taken on his personal fortune and asked de Carcavi if he knew of some mathematical explanation of the rules. De Carcavi had heard of the recently discovered century-old Cardano manuscript and relayed the possibility that the mathematics of the *Liber de Ludo Aleae* might give a clue.[6] Feeling that there must be more definite mathematical explanations of chance events, both men quickly consulted their good friend, the mathematician Blaise Pascal.

Two questions influenced the Cardano manuscript. One asked for the smallest number of times a person must throw a pair of dice to have a better-than-even chance of getting a double six. The other, known as the *problem of points*, asked for the number of points that should be awarded to each of two players in a game of dice if the game is left unfinished.[7] This may have been an old problem dating back to the late fourteenth century, but most sources claim it originated in the year 1494 when the Franciscan Fra Luca Paccioli published it in his *Summa*.[8] It came down to computing the probability of each player winning when each had a given number of turns left to play, a difficult problem that de Méré would have been aware of from his school's arithmetic texts. Such texts routinely presented questions in the spirit of insightful Arabic inheritance puzzles: A nobleman, who is fond of watching ball games, offers four ducats (about $450 in today's money) to the winner of the first eight games. Two of his farmhands play, but after one won five games and the other three, the ball was lost and the games could not be finished. The puzzle then asked: how should the prize be divided?

Pascal obtained a copy of Cardano's manuscript and examined it for signs of a solution to the dice problem, but he was skeptical of its results and wrote to his friend the lawyer and mathematician Pierre

Fermat about the problem.[9] Pascal became ill and, confined to his bed during the spring and summer, corresponded with Fermat and worked to extend the problem of points to more than two players. He felt that in a fair interrupted game of dice each player should be awarded points according to the likelihood of his or her potential in winning. Working out the mathematics of the prizes, he proved that the odds are slightly less than even that double six would turn up on twenty-four throws of a pair of dice and slightly more than even that double six would turn up on twenty-five throws.

Pascal knew that snake eyes and boxcars (double sixes) very rarely turn up since they have a 1 in 36 chance of turning up whereas seven has a 1 in 6 chance (see figure 2.1). He understood that it would be easier to calculate the probability of not throwing a double six. That would be 35/36. He also understood that the probability of two independent events happening is the product of the probabilities of the individual events and that, therefore, the probability of *not* throwing a double six on n throws is $(35/36)^n$. He calculated $(35/36)^{24}$ to be 0.509 and $(35/36)^{25}$ to be 0.494 to conclude that there was a slightly smaller than even chance of getting double sixes on 24 rolls of the dice and a slightly better-than-even chance with 25 rolls.

It was this dice problem that gave Pascal the inspiration for working with Fermat on the foundations of a theory of probability, perhaps his greatest contribution to mathematics and certainly a decisive leap forward for the field. However, by the age of twenty-five, though having a revered reputation in mathematics and physics, Pascal suddenly abandoned his work to devote the short remainder of his life to philosophy and religion.

Then, in 1657, Christiaan Huygens published the first book on probability, *De Rationciniis in Ludo Aleae* (On reasoning in games of chance), which remained the chief text on probability until 1708 when the French mathematician Pierre Rémond de Montmort published his *Essai d'analyse sur les jeux de hazard* (Analytical essay on games of chance).

But a short essay Pascal wrote in that prominent year of 1654 was not published until eight years after Huygens's probability book. In it appears the triangular arrangement of numbers that bears his name. We now know that that particular arrangement is embodied in a huge number of mathematical formulas in almost every field of mathematics, including those that are remote from the mathematics of chance. However, when it comes to chance itself, the arrangement is at the kernel.

Take a minor break to ponder Pascal's triangle, which is a triangular arrangement of numbers.

$$
\begin{array}{ccccccccccccc}
 & & & & & & 1 & & & & & & \\
 & & & & & 1 & & 1 & & & & & \\
 & & & & 1 & & 2 & & 1 & & & & \\
 & & & 1 & & 3 & & 3 & & 1 & & & \\
 & & 1 & & 4 & & 6 & & 4 & & 1 & & \\
 & 1 & & 5 & & 10 & & 10 & & 5 & & 1 & \\
1 & & 6 & & 15 & & 20 & & 15 & & 6 & & 1 \\
 & & & & & & \cdots & & & & & &
\end{array}
$$

(8)

Beyond the clear symmetry, we see that each number is the sum of the two numbers on the line above it; for example, the third number (10) on the sixth line from the top is the sum of the 4 and 6 on the fifth line. So it is easy to construct any line of these numbers. The magic here is that these are the same numbers we see when we expand the powers of a sum of two variables, say, p and q.

The constants in the expansions of the binomials $(p + q)^n$ are exactly the numbers in Pascal's triangle.

$$(p + q)^0 = 1$$
$$(p + q)^1 = 1p^1q^0 + 1p^0q^1$$
$$(p + q)^2 = 1p^2q^0 + 2p^1q^1 + 1p^0q^2$$
$$(p + q)^3 = 1p^3q^0 + 3p^2q^1 + 3p^1q^2 + 1p^0q^3$$
$$(p + q)^4 = 1p^4q^0 + 4p^3q^1 + 6p^2q^2 + 4p^1q^3 + 1p^0q^4$$
$$(p + q)^5 = 1p^5q^0 + 5p^4q^1 + 10p^3q^2 + 10p^2q^3 + 5p^1q^4 + p^0q^5$$
$$(p + q)^6 = 1p^6q^0 + 6p^5q^1 + 15p^4q^2 + 20p^3q^3 + 15p^2q^4 + 6p^1q^5 + p^0q^6$$

$$\cdots$$

FIGURE 2.4. The Pascal triangle from title page of Petrus Apianus's book of arithmetic published in 1531. D. E. Smith, *History of Mathematics* (New York: Dover, 1958), 508.

Suppose we label the *k-th* term (from the left) of the *n-th* line of Pascal's triangle as C_k^n. For example, $C_3^5 = 6$. It can be shown that

$$C_k^n = \frac{n!}{k!(n-k)!}.$$

(No need to know why this is true at this moment. Just recall that $n!$ is a symbol that means the product of all integers from 1 to n.)

There are many remarkable things to say about this term C_k^n. For now, all we want is to know that it gives us the number of different ways we can choose k objects from a set of n objects without regard to order so we can make the connection with the second column in table 8.1 (p. 103), which shows the number of ways a win can occur in four rounds of a game of chance.

The most remarkable insight into combinatorics came in 1705. In that year Jacob Bernoulli died, leaving behind reams of incomplete and unpublished manuscripts. For the next eight years Nicholas Bernoulli worked through his uncle's papers and finally published Jacob's *Ars Conjectandi* (The art of conjecture), a groundbreaking work that even today is recognized as presenting some of the most critical early notions of the mathematical theory of probability. Much of what is in the *Ars Conjectandi* had been known, but there were several remarkable additions that made significant contributions to the theory of probability.[10] There is where we first encounter the *weak law of large numbers*, an early version of the law mentioned earlier, a mathematical law that gave immense yet straightforward insight into how chance behaves in the real world, a theorem that Bernoulli proudly announced as hard, original, and so excellent that it gave dignity to all parts of his treatise (see chapter 9 or appendix C).

With this new mathematics came the trust that luck could be quantified. And from that trust came the skilled gamblers and the professional gamblers for hire. The writer Oliver Goldsmith tells the story of Richard Nash, the so-called Master of Ceremonies at Bath, a professional, an extraordinary character, and a benevolent dandy. Now, by what Goldsmith tells us, *The Life of Richard Nash* is a true biography of an honest and upright man who died in February 1761. What is of interest to us here is that this man Nash was for hire, a concept that would, one hundred and fifty years later, be spotted in Dostoyevsky's *The Gambler*. We are told that Bath was a place with few laws against gambling, though the gaming houses were responsible for the ruin of many fortunes.[11] Bath was *the* place where people of distinction went to gamble and be amused. And, as in any such place, sharps and other adventurers would be skulking in the shadows.

It was a time when slave trading was on the rise and wars involving countries all over Europe continued over colonies, trade and sea power, science, art, and literature, and practical inventions were about to explode in the Age of Enlightenment. A rising middle class was becoming informed and beginning to think and discuss politics; with its increased leisure came a desire for entertainment to fill it. And, though illegal, so came the gambling orgies in private houses, cockfight wagers in the back alleys, horse racing and increased cheating by growing troupes of professional gamesters and cardsharps turning innocent social pastime sports into a cunning, crooked business. Gambling had its grip on all ranks of society and both genders. Our English word *crook* (as in thief) comes from that era when crooked dice were used by gambling cheaters.

Unless widowed, women were still regarded as a father's or a husband's possessions and still required a chaperone to appear in public places. Yet genteel women were becoming social gamblers and by the end of the middle of the eighteenth century they were so taken with games and gambling that the papers were filled with accounts of women's wagers and ruin. The nineteenth-century British cultural historian John Ashton tells the story of Miss Frances Braddock, daughter of a distinguished army officer. She was beautiful, elegant, accomplished, and with brilliant wit, and, for all this, admired. Her father died in the American War of Independence and left her an inheritance of twelve thousand pounds, a considerable sum (approximately $1.4 million in today's money). Four years later her sister died, leaving Frances another twelve thousand pounds. She gambled her entire fortune away within a month. Penniless, she robed herself in white, and then, tying gold and silver girdles together, hanged herself.[12]

The records of that period are filled with maidens and dowagers, as well as respectable married ladies, who somehow managed to spend time at the illegal gaming houses emptying their jewelry cases, stretching their credit, and compromising their honor, all in dire hopes of recovering losses. One reporter for *The Guardian* wrote of a friend who tearfully complained that his wife spent late hours at cards throwing away his estate.[13]

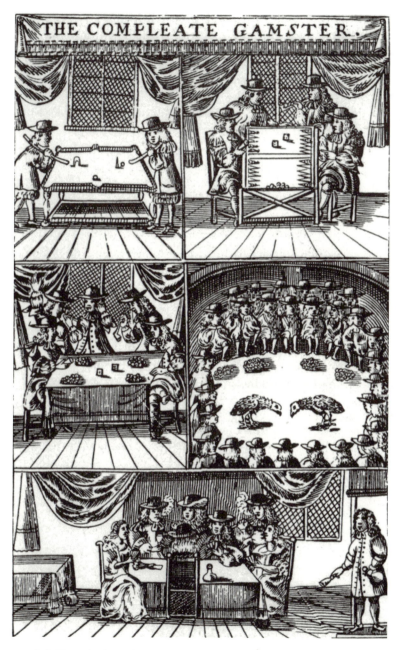

FIGURE 2.5. Frontispiece from the third edition of Charles Cotton, *The Compleat Gamester: Or Instructions How to Play at All Manner of Usual and Most Genteel Games* (1709; rept., Barre, MA: Imprint Society, 1970).

It's likely that policemen—at least high-ranking policemen—played at the gambling houses, for there was hardly any interfering, and, though dozens of illegal establishments remained brazenly open for business at all hours of day and night and high-ranking gambling parties circulated from private house to private house, there were very few arrests and fines. When scuffles broke out, the police were obliged to meddle.

Illegal gambling parties were occasionally announced in the *Times* like the news of society balls. The English poet Charles Cotton describes an early seventeenth-century gambling house, suggesting that it is a common picture.

> [E]very night, almost, some one or other, who, either heated with Wine, or made cholerick with the loss of his Money, raises a quarrel, swords are drawn, box and candlesticks thrown at one another's heads. Tables overthrown, and all the House in such a Garboyl, that it is the perfect type of Hell. Happy is the man now that can make the frame of a Table of Chimney corner his Sanctuary; and, if any are so fortunate as to get to the Stair head, they will rather hazard the breaking of their own necks, than have their souls pushed out of their bodies in the dark by they know not whom.[14]

Entertainment gamblers lost all control of reason, and so gambling became so widespread that big business saw easy profits through calculated odds in its favor. Small establishments grew with the ballooning addictions of both rich and poor. By the late seventeenth century gaming houses were no longer small backroom parlors.

♣ CHAPTER 3 ♣

From Coffee Houses to Casinos

Gaming Becomes Big Business

It is calculated that a clever child, by its cards, and its
novels, may pay for its own education.
—*London Times, November 2, 1797*

As with every other culture of Europe during the reign of
Louis XIV, the French were obsessed with gambling. The royal
apartments of Versailles were turned into casinos nightly, and in
Paris at least ten *maisons de jeux* were licensed to operate as long as
the amusements were games of skill.[1] However, most often they were
not. Gambling was embedded in court life in the late eighteenth century. So by the time of the storming of the Bastille, there were over a
hundred illegal gambling rooms in Paris. The gambling passion had
moved from monarchical Versailles to the flourishing, bourgeois
Palais Royal, where gaming clubs hummed from noon till midnight.[2]

At the beginning of the eighteenth century there were more than
two thousand coffeehouses in London, distinguished by the profes-
sions, trades, classes, and party affiliations of their clientele—lawyers
at Nando's or the Grecian, businessmen at Garrawa's or Jonathan's,
parsons at Truby's or Child's, Whigs at the St. James, Tories at the
Cocoa-tree, Scotsmen at Forest's, Frenchmen at Gile's, the leading
wits at Will's, and the gamesters at White's, a coffeehouse that was to

become the famous White's gentleman's club that still exists today. (London's population in the early 1700s was about the same as that of present-day Boston. It's worth noting that at the time of this writing there were only fifty-three independent coffeehouses and fifty-five Starbucks in the city of Boston.)

The word *club* has come down to us from the twelfth- to fifteenth-century middle English word *clubbe*, meaning *bond*, but the word as we know it came into common use in the late seventeenth century when coffeehouses—the respectable nonalcoholic alternative to taverns—became the popular meeting grounds for intellectuals and political junkies.[3] The first coffeehouse in London was in St. Michael's Alley. It opened in 1652 and was owned by a coachman to a Turkish merchant who sold the new drink Kauphy, imported from Turkey. The drink became popular and so coffeehouses sprang up all over London—Hains' in Birchin Lane and Farr's on Fleet Street.[4] These coffeehouses were convenient places for political gossip and business meetings. All ranks of life, except the lowest, were represented. Men met to talk politics, to discuss plays and poems, and to gamble with dice and cards.[5] The proprietors of these coffeehouses discovered that if they set aside special rooms for regular and important customers they could charge a membership. This is how the most famous coffeehouses became clubs—Tom's, the Cocoa-tree, and White's.

Most known and most fashionable was White's Chocolate House at 37 St. James Street, which opened in 1698—a very exclusive club frequented by dukes, earls, and members of Parliament, where red-uniformed wigged servants kept supplies of wine and chocolates going to evenly inebriate and stimulate the clientele by innocent sips and bites.[6] Word was that gambling was healthy for the spirit; even physicians prescribed an evening at White's to their patients as an entertaining diversion.[7] White's popularity grew enormously in the first twenty years of the eighteenth century, especially among the nobility and leisure class, when choice meats and liquors were served.

The betting would continue from night to morning when winners and losers would stagger home too drunk to not bet on anything that came before them. Alexander Pope warned of the deterioration of

FIGURE 3.1. Caricature of gambling at an early roulette table, ca. 1800.
http://en.wikipedia.org/wiki/Image:Gambling-ca-1800.jpg.

English society in 1728 through his literary satire *The Dunciad*, which pointed to White's and the goddess Dullness, whose mission is to bring decay, stupidity, and bad taste to England.[8]

Sometime between the French Revolution and the Napoleonic Wars (probably in the 1790s, though no one knows for sure), the roulette wheel found its way into the gaming halls of France. Unlike cards and dice, it was a game that required a croupier who would collect and pay bets—no one but the croupier touched the object that determined the chance outcome.

By the time the Napoleonic Wars were in full swing, Goya was painting his *Two Majas*, Lord Byron was writing his first poems, Hayden was completing his *Seasons* oratorio, and Beethoven his *Eroica* symphony. Great, grand gestures were all around. Merchants and consumers were becoming comfortable with the new idea of banknotes; they were far easier to carry in a purse than heavy gold coins, and so—as credit cards made purchases easier two centuries later—the new paper money was convenient. The gambler could carry more to lose. This was also the time of stock investing and speculative ventures, when gambling was simply viewed as business transactions.[9]

The French gambling mania continued even after the revolution. Napoleon legalized the important gambling clubs, but in 1837 a new act of prohibition closed the clubs and the National Guard had to be called out to evict the mobs of gamblers who refused to leave the tables. For almost half a century, the Palais Royal was the most captivating gambling attraction of Europe.[10] However, the French, perhaps as a culture, were more aware than the English of the dangers of addiction.

By the first half of the nineteenth century, gaming was a rage in England and especially in London. With increased wealth from business and factory ownerships, there were many more men and women of leisure; what were they to do with all of their time? Yes, there were affairs of their estates and households to deal with, parties and balls to organize, shooting sports and fox hunting to attend. But the lure of London clubs and society was hard to pass up.[11] Almost everywhere in London, and even in the countryside, there were young men and boys standing about on corners, in back alleys, at the wharves by the Thames, pitching halfpennies and betting on whatever stakes they could afford or not afford. Walking the streets between Charring Cross and St. James Square, one could see gaming houses in abundance. Through their open doors one could peer into the tobacco-smelling rooms to see noblemen, gentlemen, and officers wagering alongside clergymen, tradesmen, and clerks. It was as if the entire neighborhood was one enormous casino founded with the help of lawful bookmaking.[12] This was a consequence of European social reshaping that followed the industrial revolution with a relatively large leisure class. Inns and taverns offered meals and cards, and some notable private houses turned into private clubs. Gambling was easily accessible.

By today's impressions of gambling culture, it may be hard to imagine scenes of nightly gains and losses in millions of dollars, but mid-eighteenth-century gambling was a mania for many of the children of rich London gentlemen. At Almack's in Pall Mall, established principally as a gaming house in 1764, the stakes were high; every evening young men could be seen losing between ten and twenty thousand pounds.[13] These were astounding amounts, between $1 and $2 million in today's money!

Many gaming establishments were showplaces of comfort and extravagance, where mahogany-paneled walls adorned with mirrors and silk drapes met high gilded ceilings. At the Wellington, for example, the duke himself would sit where he could gamble with select company and drop as much as a staggering hundred thousand pounds in one evening of whist.

They may have been extravagant showplaces, but they were loud and crowded with addicts entranced in illusions of luck. William Hogarth depicted the typical scene in eight satirical engravings (*A Rake's Progress*) made in 1735 that tell the story of the fictional character Tom Rakewell. For Hogarth, the character was a symbol of the useless and destructive character of Britain's ruling classes. Tom comes into money by marrying an old hag with a fortune, but falls to womanizing and gambling and loses his fortune twice to end in debtor's prison where he becomes mad. In the last engraving we see Tom in Bedlam, chained and half-naked, fully destroyed.[14] In figure 3.2, we see Tom at the moment of his second loss of fortune. He kneels; his right arm outstretched with a fist toward heaven, his wig torn from his head. Around the tables we see joy and suffering from winners and losers. At the table on the left, a moneylender is writing a note to a nobleman. All but three people in the room are oblivious that the room is on fire.

Exaggerated? Maybe. However, we do know that in more than a few of the gaming houses between St. James Square and Charring Cross, many gentlemen, officers, clergy, tradesmen, clerks, and apprentices were ruined, driven to crimes that led to the gallows, even to suicide.[15]

There is a list of five hundred names of Londoners who went to their ruin in 1820—noblemen, gentlemen, officers of the army and navy, clergymen, clerks, grocers, horse dealers, linen drapers, masons, and booksellers.[16] The most well-known is George "Beau" Brummell, the legendary dandy and fashion fanatic who took four hours each day to dress. The fastidious man was reputed to have simplified men's fashions down to the plain business suit, shirt, and tie. He was a gentleman who allegedly polished his boots with champagne, a close friend of the prince regent George IV, and a man who

FIGURE 3.2. *Scene in a Gaming House (A Rake's Progress)*, 1735. William Hogarth (1697–1764). The sixth in a series of eight engravings illustrating Tom Rakewell's second loss of fortune. From Sean Shesgreen, *Engravings by Hogarth* (New York, Dover Press, 1973), plate 31.

had such a severe addiction to gambling that he would wager on anything from Napoleon's triumphs to his own death. Brummell was the quintessential addict. He frequented White's. His debts forced him into disgrace, into quarrels with the prince, and eventually into fleeing England for Calais, where he lived in poverty and squalor for the remaining sixteen years of his life before dying of syphilis in an asylum in Caen.

However, by 1870, gambling in England had almost ceased, or at least quieted down to a few clandestine table operations here and there, slinking and skulking away into corners and holes, as a consequence of the prohibition by the New Gaming Act of 1845.[17]

Statutes were already in place to imprison anyone designated a rogue or vagabond and to subject such designates to hard labor and corporal punishment. The Unlawful Games Act (which was passed as far back as 1541 under Henry VIII) merely discouraged gambling without enforcement. It was not so much directed at the morals of those who gambled as it was aimed at the children of noblemen who were recklessly depleting family fortunes and inheritances.[18]

After the New Gaming Act, gambling activity in England did abate. It may have had less to do with the prohibition and more to do with the expansion of other leisure opportunities. Travel by rail and ship had become easier; museums, theaters, opera houses, and concert halls were built to accommodate a new demand for the arts, and weed-overgrown fields were converted to public parks where people could stroll and talk.

While some laws in France were more relaxed by the end of the nineteenth century, private gambling houses were illegal, the same as in Prussia and Austria. In Bavaria, the law distinguished between games of skill, which were legal, and games of luck, which were illegal. However, in Spain, all wagers were lawful.

After so many fortunes were lost in the Paris casinos, French law decided that prohibition was the best thing, and so on the last day of 1837 the Paris casinos closed. With widespread prohibition in England and France, affluent gamblers traveled for the cure. From May to October they went to the baths, the thermal baths of Spa, Baden-Baden, Homburg, Wiesbaden, and dozens of other reconstructed Roman bath towns in central Europe. For some reason, these establishments escaped inspection of the authorities. A *New York Times* article written by a Paris correspondent claimed on August 30, 1858, that all Paris fled to Baden-Baden and other Prussian resorts for the summer.[19]

And why not? Draw a straight line from Paris to Vienna and Baden-Baden will be at the midpoint, just past the Rhine. Travel from Paris in 1858 would have taken two days by rail and perhaps as many as five by coach. For the Russians, such as the Turgenevs and Dostoyevskys, coming from St. Petersburg or Moscow, it would have taken weeks in horse carriages, bouncing over rutty roads, stopping for food, drink, and sleep in inns and taverns, though some of it in the beautiful countryside and resorts along the Baltic Sea. Once there, they could alternate their time between the relaxing warm mineral baths of the pump house and the intense excitement of roulette and other games in the posh, golden-pillared, glass-domed casino where they would be among other foreigners from all over Europe who must have known about the mathematical house advantage but likely didn't care.[20]

These resort casinos were the instruments of a new kind of gaming industry where the mathematics was so heavily biased toward the casino profits that many fortunes were quickly lost and never regained. After heavy losses from trente-et-quarante at Baden-Baden, the French socialist politician Henri Rochefort wrote,

> If the public were not such simpletons . . . people would make the mental comment that the luxury flaunted in these casinos, the percentage paid to the political authorities who tolerate them, and the expensive theatrical performances given in them are proof positive that the player has not the slightest chance of winning a single penny.[21]

And yet, by 1872, the time Prussia got around to its gambling prohibition, the Baden-Baden resort was a sanctuary of fashion, fortune, and flirtation, consistently attracting over fifty thousand visitors a year from all over Europe. Here, in this place of class indifference, was an uncanny mixture of both the cheerful and miserable in a paradise of tall pines over groomed gardens, baths, splendid restaurants, and posh gaming rooms where a duke, a prince, or even an incognito king would now and then be spotted.[22] We know more about the literary crowds, for they were happy to write about themselves. Gogol, Turgenev, Dostoyevsky, and Tolstoy mixed with other members of the intelligentsia and aristocracy as they frequented the resorts of Ems, Homburg, Wiesbaden, Aachen, and Baden-Baden, though some Russian aristocrats went for the baths and social environment and were not so obsessed with gambling.

However, in 1872, by a bill of the Chamber of Deputies at Berlin reasoning that gambling houses brought shame and dishonor to Germany, all those glittering resorts closed and the gambling crowds moved south to Monaco, which was by then established as the jewel of the Riviera because of its rich elegance.[23] The railroads were running along the coast from Nice, the roads over the once impassable terrain of the Alps from the north were paved, so the crowds could pack into Monte Carlo at all seasons of the year. The casino, not as elegant as it would soon become, attracted a lower class of gamblers along with nobility, such as the Prince of Wales (the future Edward VII under the pseudonym of "Captain White" or "Baron Renfew"). But this new casino was different; by the 1920s it would be the gambling mecca of the world, housing a theater, highlighting famous actors such as Sarah Bernhardt as well as the Russian Ballet, an opera house heralding Enrico Caruso, chic dining and lodging, sailing and automobile races. Monaco, with its phenomenal tax advantage (meaning no income tax at all), attracted the wealthiest people of the world to be its new citizens and residents.

♣ CHAPTER 4 ♣

There's No Stopping It Now

From Bans to Bookies

[T]he sweetness of winning much
and seeing others lose had turned to the
sourness of losing much and seeing others win.
—*George Eliot, Daniel Deronda*

Traditional American entertainment and relaxation with dice and card playing goes far back to the original colonies, over a hundred and fifty years before the American Revolution. But the explosion of gambling was inevitable under the new democracy. And with the purchase of Louisiana at the turn of the nineteenth century came New Orleans, where cards were dealt, dice were rolled, and craps (the American version of hazard) was played day and night. For the next hundred and fifty years that city would remain a sanctuary from the country's gambling prohibition.

Just as in most of Europe, America had laws against gambling; yet in the bigger cities, in the more fashionable districts, gambling houses were plentiful. Gambling floated up the Mississippi to Vicksburg, then on to Memphis, branching up the Ohio to Louisville, Cincinnati, and Cleveland. By the 1830s paddle-wheel riverboats were sailing upstream along the Mississippi, attracting an eclectic group, from wealthy southern plantation owners with more money than sense to snake oil salesmen and cowboys.

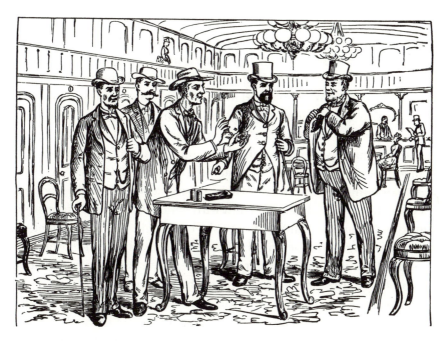

FIGURE 4.1. Riverboat gambler sharking a wealthy businessman in three-card monte. From George H. Devol, *Forty Years a Gambler on the Mississippi* (Cincinnati: Devol and Haines, 1887), 193.

The typical gambler knew almost nothing of the mathematics of risk but did have a sense of the rarity of hands—that drawing a full house is almost a hundred times more likely than drawing a straight flush and that a straight is almost twice as likely as a flush. Such information was part of the culture, though it wouldn't have taken much mathematics to figure it out (see chapter 10, in particular table 10.1). Riverboat gamblers wagered on the wildest contests from craps to fly loo, betting that a fly in the room will land on one sugar cube or another at the end of a long table. Rich southerners were seen slapping papers on the table and crying out in southern drawl, "The deed to my plantation!" One of the richest plantation owners of the Louisiana lost everything, including his plantation, and became a croupier at a New Orleans gambling house.[1]

In 1848, when the first gold searchers started arriving, the city of San Francisco saw a rise in gambling. But when the next waves came late in the year to that city of only 25,000 at alarming rates (over 40,000 by the end of the year) from South America, Australia, and the East Coast cities of the United States, gambling became a serious concern. Saloons turned into gambling dens crowded day and night with immigrants mad for excitement and testing their luck on what fifty cents could do with the turning of a card. These new immigrants had some pocket money. They sold their land and possessions to move west in search of fortune. No doubt, where there is someone with money, there is someone ready to take it away. And so new gambling houses and brothels, posing as dance halls, opened all over the city, ready to entertain with wine, attractive women, and illusions of fortunes. In San Francisco one could play faro, chuck-a-luck, brag, high/low, or grand hazard in almost any of the over one thousand dodgy dance halls at all hours, day or night. New Orleans was no longer the gambling capital of the country.

During the Civil War, cities and towns rescinded gambling prohibition; gambling spread to all corners of the continent bringing new games such as monte and euchre. Card manufacturers designed patriotic face cards for each side, the Union and the Confederacy, and made a fortune. Each side saw this as a way for troops to learn more about the enemy and it would become a military tool to be used again in World War II (with the silhouettes of German and Japanese fighters), the Korean War, and the Iraq War.[2] During the 2003 U.S. invasion of Iraq, U.S. troops were given playing cards with fifty-two of the most-wanted enemies, ranked by importance—Saddam was the ace of spades, Qusay was the ace of clubs, and Uday the ace of hearts.

In downtown New York, at the end of the nineteenth century, there was hardly a street without a private gambling house well-known to the police, who would profit from kickbacks as well as a few turns of faro. Gambling in New York was the common pastime of the poorest workmen, discharged servicemen, and precocious street boys.

The center of gambling attraction in New York was at Pat Hern's place, on Broadway, near the corner of Houston Street, where visitors would be generously treated with wine and supper. Or Morrissey's in Union Square, splendidly furnished, welcoming, and, here, too, lavishly generous in food and wine at all hours of the day and night.

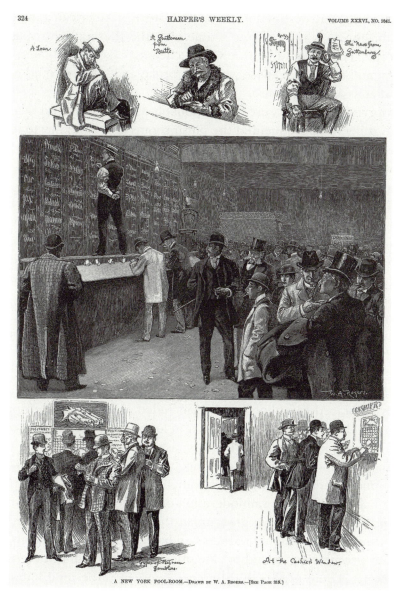

324   HARPER'S WEEKLY.   VOLUME XXXVI, NO. 1841.

A NEW YORK POOL-ROOM.—DRAWN BY W. A. ROGERS.—[SEE PAGE 319.]

FIGURE 4.2. New York City poolroom in 1892. From *Harper's Weekly*, April 2, 1892, University of Las Vegas Special Collections; image © 2000 HarpWeek, LLC. See also David Schwartz, *Roll the Bones: The History of Gambling* (New York: Gotham, 2006), 335.

Such perks were destined to become a feature of American gambling establishments, as it was for a time with the thermal bath spas in Europe. There would be free wines, cheeses, cakes, chocolates, biscuits, and nuts. Even those who were broke and hungry could get a meal at a gambling house.

Morrissey's visitors were the fashionably dressed, wild, and wealthy high rollers who frequented the plush rooms night and day to drop tens of thousands of dollars at a turn ($1,000 in 1900 would be worth more than $26,000 today). Enterprising Americans had figured out a new way to lure and detain anyone with pockets of coins.

There were also sleazy houses. At 102 The Bowery, there was a downscale gambling house, where those who lost everything at Morrissey's could win a few bucks and return to lose again at Morrissey's. If you passed at night, a street hawker would size you up and, depending on what he saw, yell either "Hallo, old sport come and try your luck—you look lucky this evening; and if you make a good run you may sport a gold watch and chain, and a velvet vest, like myself" or "You look down at the mouth tonight! Come along and have a turn—and never mind your supper tonight."[3]

As far back as the Revolutionary War, Americans had enthusiastically embraced gambling and had included almost every European game in their repertoire from poker to lottery.[4] (The original idea of the lottery goes back to the Han Dynasty of China when funds were raised by a lottery system to finance the Great Wall.) Almost from the time the idea first came to colonial America, the lottery inspiration exploded into an immoderation of private lotteries and raffles administered not by the government to augment tax revenues for building roads or waging wars, as had been the traditional sponsors, but by private citizens and private organizations to build their bank accounts. There is a small difference between raffles and lotteries. In a raffle, each participant buys a ticket for a random drawing of a fixed prize, whereas a lottery prize grows with the number of ticket purchasers. Raffles go back to first-century Rome when members of the Forum would throw prize parties, banquets, and Saturnalias for political supporters where guests received door prizes. However, later, Florentine and Genoan merchants inflated the idea by encouraging the sale of products and used it to generate private profits.

Still later, Venetian princes saw it as a means to generate considerable municipal revenue. Some say that the first recorded lottery took place in 1444 when the Flemish city of L'Écluse raised funds for repairing the surrounding city walls.[5] However, that was more of a raffle than a true lottery, for the prize was fixed beforehand at 300 florins (about $300,000 in today's money). That was a time when an unskilled laborer earned about a penny a day on top of room and board, a time when a whole chicken or a night at a nice clean inn cost about one English penny. In today's state lotteries it is often the case that nobody wins and so the pot grows and becomes more attractive for the next round and the next pool of participants. Though we must recognize that some form of lottery came to the Italian colonies by way of Flanders, the first true form of lottery happened in fifteenth-century Venice. By the mid-sixteenth century, Venice was teeming with lotteries—the Rialto district was filled with lottery hawkers selling tickets for cash prizes of 1,500 ducats (about $170,000 in today's money).[6] This new form of gambling flourished in the eighteenth century to finance wars, museums, and even universities.[7] (Lotteries were used to finance the American Revolutionary War, the rebuilding of Faneuil Hall in Boston, the establishment of the British Museum, and the building of Harvard, Yale, Princeton, and Columbia.)

The lottery came to America by inheritance and came to be part of the everyday life of the colonies; it played well in the parent country and when colonists were faced with dire economic problems, they naturally turned to what had worked before.[8] These were still private lotteries, administered by companies claiming social benefits. In the thirty years before the Revolutionary War, 161 lotteries were approved by twelve colonies to build revenue for diverse purposes from building a lighthouse in New London, Connecticut, to paying for wars to founding Columbia University (at that time King's College).[9] In England at that time, alongside a rush of fraudulent lotteries that ignited disfavor among respectable folk, legitimate lotteries funded municipal projects, such as bridges and an aqueduct to serve London. After the American War of Independence, when

FIGURE 4.3. National lottery ticket, 1821, as it appears in John Samuel Ezell, *Fortune's Merry Wheel: The Lottery in America* (Cambridge, MA: Harvard University Press, 1960). Reproduced courtesy of the New-York Historical Society.

so many in America gambled, the lottery was thought to be an honest business. Even prominent and upright statesmen such as Benjamin Franklin ran lotteries.[10] George Washington bought tickets for the Virginia lottery of 1790, which was organized to help pave the streets of Alexandria. And in March 1826, Thomas Jefferson, then eighty-three and facing poverty in the last months of his life, received authorization from the Virginia legislature to dispose of his assets, including Monticello, as bonus prizes by lottery.[11] However, by the 1820s, a growing trend of lottery gambling was becoming noticeable; it began to be viewed as a threat to legitimate business and industry, and there was a growing fear that America's national industry would be gambling.[12] Fraudulent lotteries were appearing everywhere. With the potential for enormous profits, lotteries became the cheating tool of any scammer with access to a press to print raffle tickets and means to advertise his own sweepstake. Agents disappeared after the drawings, so prizewinners seldom received their due winnings. Even the contractor for the Grand National Lottery (see figure 4.3) disappeared with several hundred thousand dollars and the grand prize winner—who should have received $100,000 (roughly $1.8 million in today's money)—had to appeal to the U.S. Supreme Court to sue the City of Washington for his prize money.

Largely due to rampant lottery fraud, lotteries were made illegal in almost every state in America by the time of the Civil War; but after, with so much financial turmoil resulting from the war itself, the Southern states rescinded laws banning lotteries. In the north, the prohibition remained in effect, though it was never enforced. From 1872 to the end of 1894, thousands of lottery vendors all over New York City were operating untroubled by the law, many working from unconcealed shops, others by peripatetic dealing. As of January 1, 1894, lotteries in the United States were illegal. President Cleveland signed a bill into law closing all forms of interstate commerce to lotteries, ending the American lotteries for the next seventy years.

After World War II, most states changed their gambling laws to permit charitable organizations to conduct bingo operations and, over time, other gaming businesses, such as raffles and pull-tabs (multilayered paper tickets containing symbols hidden behind perforated tabs). Then, in 1964, because of mounting opposition to tax increases, New Hampshire legalized a low-stakes biannual lottery. New York and New Jersey followed, and the first legal off-track betting system (OTB) opened in New York. Then New Jersey reversed its gambling laws to permit casino gambling in Atlantic City. Growing liberalization of gaming sentiment led to *California v. Cabazon Band of Mission Indians,* when the U.S. Supreme Court ruled in favor of permitting gambling on Indian reservations. In 1986, the Mashantucket Pequot Tribal Nation opened a high-stakes bingo hall in Connecticut, generating $13 million in gross sales and $2.6 million in profits in 1986 dollars, which would translate to $25 million and $5 million in 2009 dollars. Soon afterward, Foxwoods Resort Casino opened within walking distance of the Tribal Nation to become the largest casino complex in the world. As of this writing, it has slipped back to being only the third largest—the second largest is the neighboring Mohegan Sun and the largest, to be called the Venetian Resort, is currently under construction in Macau. It will become the Asian capital of gambling as well as the world's largest casino. In 2009, at the time of this writing, New York State is thinking that it has the answer to its budget deficit by opening development rights to Aqueduct, the state-owned racetrack, to bidders who will pay $250 million and add 4,500 video slot machines.[13] And so, the cycle of gambling continues.

Perhaps, as Edmund Burke put it in a speech to the House of Commons at the end of the eighteenth century, "Gaming is a principle inherent in human nature."[14] And perhaps gaming is natural for the survival of our civilization. The biggest gaming houses in the world are not the casinos of Atlantic City or Nevada but the stock exchanges in New York, London, Frankfurt, Tokyo, and Hong Kong, and 115 other stock exchanges around the world, which every working day shift billions of securities from one place to another in adrenaline-gushing ventures. These gambles that may be primed by mixtures of greed and profit support our civilized existence, without which goods and services for much of the growing population of the developed world (now approaching seven billion) would starve.

Sometime back in the twelfth century a group of Frenchmen came up with the idea of selling shares in the ownership of their textile mill in order to finance its ongoing operations. In 1553, some rich and daring London businessmen gambled by purchasing shares in the Mysterie and Compagnie of the Merchant Adventurers for the Discoverie of Regions, Dominions, Islands and Places Unknown. The company sent three ships under the command of Sir Hugh Willoughby to places unknown at a time when news from sea and far-away lands would take months, if ever, to return. Early news was very bad; two of the company ships along with their crews froze in Arctic ice and for a time it seemed that the shareholders had lost their bet. But the third ship made it all the way to Russia and the crew negotiated a lucrative trade treaty with Ivan the Terrible, opening trade between England and a very rich Russia. The original investors hit a jackpot.[15] Other companies quickly saw share selling to independent investors as a business model for success.

And so the idea of buying and selling shares spread to other countries in Europe and was eventually picked up by governments that sought to finance their wars. But several centuries would pass before the first real commodities trading stock exchange opened in Antwerp, Belgium, in 1531. Hamburg picked up the idea in 1558, followed by Amsterdam in 1619, followed by London and Paris later in the century. In 1792 Alexander Hamilton (as secretary of the

Treasury) advised that American government debt securities be traded on the corner of Wall and Broad streets in New York City, where neighborhood coffee shops were abuzz with traders of bonds and stock issues. Twenty-three years later, the trading moved indoors at 40 Wall Street and later became the New York Stock Exchange.

At first, the exchanges were simply bets on current and future values of commodities with guesses about what the world would bring to the marketplace in the way of metals, textiles, and agricultural products. In Amsterdam smart money was laid on the Dutch East India Company (the first company to issue stocks and bonds). But who could know for sure that it would be the Dutch rather than the Portuguese who would secure trading posts in the Spice Islands? The bets were on fleets returning with everything Eastern, from cloves to gems. However, there was always a **better-than-even** chance for a shipwreck.

Take the case of the Dutch tulip mania that moved through buying and trading in 1636 when prices went so high a single rare tulip bulb could fetch a house and land, as well as a carriage with a pair of horses.[16] The next year the tulip market got jittery and the market wilted—imagine the pain of those who traded their houses for valueless bulbs. Or take that of the South Sea Company of London, whose shares in 1720 shot up tenfold in just eight months. Thousands of investors were bankrupted by betting on the company's overvalued claims of profitability; in turn many investors, including the king and prime minister, put down very little of their own money to buy the shares on credit and the company was left worthless.

And what about insurance? Is that not a form of wagering with a hefty hedge? The idea was not born in the coffeehouses of London but did get a terrific boost from Lloyd's, the coffeehouse at the corner of Abchurch Lane and Lombard Street owned by Edward Lloyd. There, a group of underwriters met to gamble on everything from ship risks, overdue voyages, fires, burglaries, earthquakes, and tramcar accidents to insurance against having twins or against the queen passing through certain streets on her Jubilee. Shipowners and merchants frequented Lloyd's, and, since Lloyd himself was a newspaper publisher, his establishment naturally became the hub of shipping news and bets on commerce. (Until the end of the nineteenth century, almost all newspaper news was about shipping, listing ships' arrivals and their contents.) The company grew to become the world's leading insurance market. Even today, it is not a company but rather a market for insurance, just as it was when its underwriters met at the coffeehouse in the second half of the seventeenth century. A company called Phoenix that met at the Rainbow Coffee House at 15 Fleet Street underwrote the first fire insurance policies sixteen years after the Great Fire of London.[17] Was that not gambling? Those early coffeehouses, though dim and scant, were perfect places for betting and insurance transactions to occur along with merchant affairs and all kinds of political gossip, and both social and business news.

In the 1920s shareholders borrowed to invest in a market, but their investment returns could not cover the interest rate on their borrowed money. When prices fell in 1929, millions of investors couldn't pay their loans and the banks were left with too many defaults to pay depositors who demanded their money.

At the age of fourteen in 1891, Jesse Livermore had a job chalking stock quotes on a board for Paine Webber. By recording and analyzing stock patterns over the next six years, he was able to make some predictions and make $10,000 from $5 by trading at bucket shops, small-time betting shops that would accept small transactions, often within coffeehouses, drugstores, and hotels. Some were dishonest, posting false prices, but most were legitimate. Livermore made his first million just before a substantial crash in 1907 and, in the wake of the 1929 Wall Street crash, sat on a reputed fortune of $100 million.[18] How did he do it? By mastering the art of selling short; in other words, by betting that the values of some stocks would drop, and even betting on the likelihood that some companies would go bankrupt.[85] Very clever, but by the age of sixty-two he was bankrupt.[19] A year later, on November 28, 1940, at 4:30, Jesse Livermore walked into the Sherry-Netherland Hotel on Fifth Avenue, sat at the end of the cloakroom, and shot himself in the head with a .32-caliber Colt automatic pistol.[20]

At the end of World War II, Alfred Winslow Jones, who was then a financial journalist for *Fortune*, took Livermore's short-selling idea one step further in creating a fund that hedged investment risk, which has now come to be known as a *hedge fund*.[21] As the name implies, he minimized his risk by hedging his bets, a phrase that came out of roulette where one places a bet *on the hedge* between two competing choices.[22] Until then, most investments were bought *long*, that is, in the hope that their values would soon increase. The idea was as old as insurance; in England's great lottery years of the early nineteenth century, Londoners could insure their lottery tickets by betting that their number would not be drawn on a particular day.[23] But Jones hedged his bets by going *long* as well as *short*; that is, identifying shares that were overvalued and betting that their value would decrease.[24] He borrowed such shares, immediately sold them, and later when their value plummeted bought them back at the lower price, exactly the sort of thing Livermore did to make his fortune. Any profit made would cover his long shares and so his risk in picking stocks more aggressively was minimized.

Thousands of speculators, including hedge funds, buy shares in companies that have no products other than the cash that comes from issuing their own stocks. They are the penny stocks whose prices can be manipulated by "pump and dump" schemes. These are companies with little fundamental value and no other business beyond the capability of printing shares by the millions. Promoters manipulate the stock by posting glowing information on bulletin boards or through brokers who contact investors through mass e-mails. Speculators, who do not own or borrow shares, engage in *naked short selling*; they place bets that stock prices will decline. Pumping and dumping works in the short run before everyone discovers that the company is worthless. Then the promoters bail out and the price hits a bankrupting bottom.

And now, in the twenty-first century, we can bet not only on plunging values of shares but on credit-default swaps, making it possible for us to bet that particular companies will default on repaying their loans. "Such bets on credit defaults," said the billionaire George Soros, predicting the 2008 economic meltdown a year before it actually happened, "now make up a $45 trillion market that is entirely unregulated. It amounts to more than five times the total of the US government bond market. The large potential risks of such investments are not being acknowledged."[25]

Shipwrecks on the high seas are now rare and the outcomes of many ventures are more certain, but for the novice, modern trading is as much a gamble as what happens on the green tables at Vegas.

♣ CHAPTER 5 ♣

Betting with Trillions

The 2008 World Economic Calamity

> Only when the tide goes out do you discover
> who's been swimming naked.
> —*Warren Buffett*

It was October 2008. The world economy was beginning its tailspin when I interviewed George ("Jerzy") Sulimirski, a champion backgammon player on the world circuit, in his London home. He had just confessed to me that he once lost a game to a notorious cheater.

"Where there's money there's cheating," he said in a mellifluous trace of Polish accent. "Kind of giving providence a helping hand. If a person can take a watch off your wrist without you knowing it, as a skilled trickster can, he can effortlessly cheat you at the game board."

George lost big time at a game to a cheat who used a partially magnetic board and dice loaded with metal. The cheater could favorably control the outcome by where on the board he tossed his dice.

"The rule is," he continued, "know whom you're playing with."

He then went on to tell me about the rules and modes of the game. Backgammon has many different sets of rules and modes. One mode is called a *chouette*. In a chouette, there's the man in the box called the *box*. There's the *crew*, that is, all the other players led by one other player called the *captain*. The captain rolls the dice and makes all the final decisions on the moves. Now the captain and crew play against the box.

"A three-person game of chouette," he resumed, "is open to one of the old hustling tricks of partner splits, where two people agree to split the winnings and losses. In this scam, the unsuspecting odd man out, the box, who is not splitting his gains and losses with other players, is not aware that every time he is the crew his captain loses."

"But the captain loses too!"

"Yes, but the captain's losses are always half that of the box's. You have to remember that in all gambling there is cheating and greed."

"Greed? Tell me more about greed."

"There's nothing wrong with greed, unless it's uninhibited; nor with risk, unless it's reckless. And there's a fine line between extremes of risk behavior and recklessness."

"Don't you mean carelessness?" I asked.

"No. I mean recklessness in the spirit of the word—deliberately courting danger."

"I understand the dopamine effect of courting danger—even death—but deliberate? What kinds of games are played with a deliberate invitation for danger?" I asked, very excited about where the interview was going.

"The stock market, you know. It's the biggest casino on earth!"

"Casino?"

"Yes, casino."

Even the masters of the universe, who every moment of every day place their bets on what's ahead, did not see the coming of such a precipitous decline of the world economy. Equities markets are extremely complex; there are no algorithms, no model by equation. So investments should be viewed as simply poker games based on a risk-reward evaluation. With poker, you may compute the probabilities of being dealt a favorable card. You have to weigh the risk of not getting a favorable card with how bad it would be to lose the pot. You have to weigh the odds of your hand overriding the one you are trying to beat. The same is true with financial markets. The amount of risk you are willing to take should be compared with the return you are likely to clear. You buy and sell a stock according to appraisal and judgment, looking at past and current earnings, growth potential, competition, and so forth. But in the end, whatever the balance sheet, the shrewdest investment is still a gamble. As with any game of finance, without strong regard for human wishful thinking and other unquantifiable behavioral factors, mathematical models of risk used by financial engineering sharps may envision the action very wrongly, especially in an atmosphere of bull market profit chases. When a few small-time investors buy stocks and lose, it's no big deal. In fact, the market works best when that happens, because it encourages a modicum of volatility in pricing so others can make money. But when financial institutions buy and sell, the precipitous volume of their transactions can influence strong resonant waves that—short of government interference—could bring down the entire world economy.

Consider the case of Jérôme Kerviel, the thirty-one-year-old French trader who perpetrated the largest trading fraud in history. He brought Société Générale, one of the most admired financial institutions in the world, to a staggering net loss of 4.9 billion euros. In July 2005 he sold short €10 million of a European insurance company, gambling that its stock price would fall. There was no clear indication that the price would fall, but by Kerviel's luck the entire London FTSE fell due to nothing more than the July 7 London bombings, bringing him a net profit of half a million euros. That lucky gamble contributed to a promising history of reinforcement.[1] Kerviel told the police, "It makes you want to continue; there's a snowball effect."[2] And so his risky behavior intensified. He made covert purchases on the order of hundreds of millions of euros, which turned into further substantial profits. To conceal his purchases he conjured fictitious trades to offset and hide any gains he made. It wasn't long before he was short selling in anticipation that the global markets would suffer severely from the sub-prime mess. Selling short millions rapidly turned into selling short hundreds of millions, which very quickly turned into billions. Kerviel had gambled that the sub-prime mess would severely shake the markets further downward. It did, and by the end of 2007 his positions gained a colossal 1.5 billion euros. However, by the beginning of 2008, believing that the market had hit bottom and that recovery was inevitable, he began to boldly bet on futures, extending his exposure to nearly fifty billion euros. He was under the influence of overconfidence and the typical innocent gambler's illusion of control. Overestimating his ability, he mistook a talent for making wise investments with an ability to control pure chance. That's when things began to go badly. The equity markets were falling and an exposure of that size without a hedge could have shaken Société Générale into bankruptcy. The bank had to quickly sell off fifty billion euros' worth of futures without causing too much notice, which could have triggered panic. It took three days to do so. In the end it avoided complete catastrophe but incurred the largest single-day trading loss for a single company in banking history: 6.4 billion euros.[3]

The history of humankind is filled with risk-taking adventure. Risks are natural, the balance between expectation and fate, the human desire to wager in the clash of opposing chances. We wouldn't have banks without risk, nor would we have useful tools and products like Google or cars like the Prius. Someone was out there willing to take a risk in lending $100,000 to two Ph.D. students from Stanford working in a garage on a new type of Web search engine.[4] Toyota gambled with over a billion dollars in development money to produce its hybrid fuel-efficient car. It ran into trouble with impossible demands, serious miscalculations, and technological difficulties such as battery nightmares; however, after all the anguish, the car's reception was a major success, just in time for a market hit hard by soaring gasoline prices. It's hard to imagine now how nervous the Toyota research team must have been. Like any company deciding which research projects to fund, Toyota had taken a huge gamble. Without the risk originally taken by free-enterprise entrepreneurs, neither Google nor the Prius would have come into existence. Those investments paid off. But even if they hadn't, the attempts would have been noble risks, not reckless ventures. The banking industry's extensive risks are another story. They were reckless ventures goaded by unrestrained greed. "Be fearful when others are greedy, and be greedy when others are fearful," Warren Buffett once remarked in trying to explain his rule for buying at a time when investors were terribly fearful of the weak economy.[5]

Greed is a judicious maximizing of self-interest. It can be unsympathetic to the general good. It has always been with us and always will be. In the last century, isolated cases of risky Wall Street leverages were insulated from global effects. With economic globalization, however, almost all banks are entwined in a strong web of dealings that make them vulnerable to the behavior of one.

Though Société Générale's woes were felt across world markets, its loss was someone else's gain. In the three days, while the bank was frantically liquidating Kerviel's futures positions, hundreds of other traders were making money selling short, placing their chips on a falling market. And even when the world markets were tumbling some people were still making money. Worse yet, when Wall Street almost fully collapsed, stock traders, top-heavy corporation CEOs, and high-roller investors—those masters of the universe—found themselves in the humiliating position of asking for government handouts.[6]

By the end of the summer of 2008, the world saw just how tightly its financial institutions were tied. For more than a quarter century, the U.S. banking industry had been lending with little control through agents who would receive commissions based on the number of contracted mortgages. Risky loans were given willy-nilly to anyone with a job, no matter how little the job paid. Even more risky were the loans brokers called *ninja loans*, given to people with no job or income, and with no questions asked. People were able to borrow 100 percent of their inflated house prices with no money down. Along with mortgages, those same people and their children were bombarded with incentives to apply for credit cards with minimum balance monthly payments that would swell to unendurable amounts. During President George W. Bush's tenure at the White House, Congress and the administration did almost nothing to help the poor and middle class, there had been only two minor minimum wage increases that had not kept up with inflation, health insurance premiums soared, and the famous tax breaks to the wealthy never really trickled down to the non-wealthy. Earning power dwindled with inflation in home heating oil and automobile gas. And with inflated oil and gas came inflated prices on the masses of commodities whose production and supply were dependent on the high price of fuel, in particular food and food products.

While the rogue trader Jérôme Kerviel was bringing a single French bank to its knees with a 6.4 billion euro loss, hundreds of thousands of uncontrolled loan officers were greedily encouraging millions of unqualified people into the housing market, ignoring old-fashioned risk-to-gain ratios. Colossal as they were, Société Générale's losses were minor compared to what the global banking industry would suffer: heavy exposure to losses on mortgages and consumer debt.

But what brought down the world financial equities? High oil prices were only partly to blame. Greedy mortgage brokers left the banking industry with a colossal number of homeowners who could marginally afford the homes they heavily mortgaged at variable rates, homes that were valued less than their mortgage principals. Inflation and hence rate increases put many over the margins of their budgets. Without alternatives, they abandoned their homes and mortgage responsibilities. Loans went bad. Banks suffered such severe losses that they hesitated to make loans out of fear of making more bad loans.

Banks were still solvent. They had sufficient reserves. However, their capital eroded so much that without assurances that they could meet all their financial obligations, lending slowed to a dribble. Absorbing deteriorating mortgage-backed securities also fed bankers the scary possibility of bank insolvencies emerging from falling house prices.

Banks are in the business of lending. When they don't lend, they don't generate capital. And so naturally the entire financial cycle spiraled downward. That forceful cycle dragged in those who could still afford to pay their mortgages but were still feeling the pinch from increased prices. They cut back on spending, which reduced the profit margins of companies selling goods and services that were not absolutely necessary for comfortable survival. Because of dwindling bank capital, businesses could not borrow at acceptable rates and were forced to lay off employees. And so the spiral continued with increasing drive.

Société Générale's staggering loss was repeated at banks around the world. Citigroup suffered tens of billions in losses from the sub-prime related debt. In Britain, the government was forced to nationalize Northern Rock and Bradford & Bingley. The financial giant Bear Stearns died in the unwarranted credit panic over its large blocks of tightly invested securities; it was not able to roll over short-term loans. Fannie Mae and Freddie Mac had to be nationalized by the Treasury and the Federal Reserve. With Bear Stearns dead, the lack-of-confidence panic shifted to Lehman Brothers, which, as a result, found it difficult to borrow money.[7] So it, too, died. However, unlike Bear Stearns, which was rescued by the Federal Reserve–engineered fire sale to JP Morgan, Lehman was abandoned to bankruptcy (the biggest in U.S. history), and its creditors lost almost everything. Next was the insurance giant American International Group (AIG). It had exposed itself to mortgage-backed securities through credit-default swaps. Most analysts still agree that it was basically a sound company, yet almost overnight the markets themselves marked the company as a risk. Fear that it was next to go under drove it to be next. Though the Federal Reserve, which took an 80 percent stake, rescued AIG at an initial cost of $85 billion in a gamble, creditors were not protected and the company was left trading at a 40 percent loss. Bank of America rescued Merrill Lynch by a takeover. Then Washington Mutual's doors closed—the largest bank failure in American history. Wachovia disappeared. This was clearly enough to send the word out to uninsured depositors and millions of small investors who rely on banks and brokerage firms for day-to-day cash: GET OUT WHILE YOU CAN. And they did. They left the securities market for Treasuries and safer institutions. With fewer investors in the market, the volume of trading shrank and with decreased volume came—as it always does—increased volatility. In swift progression shaken financial institutions found themselves taking catastrophic losses. Like ice cubes in a sink of hot water, a meltdown was inevitable.

The crisis spread to emerging markets from Russia to Brazil. Fearing that a recession in China would lead to cuts in exports and hence factory closures, investors pulled their money out of the Shanghai stock market and saw it drop by 60 percent over the year. The Swiss bank UBS took a hit of $42 billion on the U.S. sub-prime write-downs and had to lay off thousands of employees. As default rates increased, creditors began to bail out from sub-prime debt, causing mortgage financing and home building to slow to a crawl—some would say, collapse. Even hedge funds suffered, because in the cycle of panic, investors began pulling their money out, forcing fund managers to sell stocks and assets far under their value.

Kerviel was not the only trader taking big risks. Others traded under cocky master-of-the-universe notions of market durability, ignoring the possibility that some small rare event may cause a global catastrophe. They gambled with the market, believing that it ran by some kind of perfectly efficient rule, when, in effect, it was no more predictable than the law of large numbers' ultimate prediction of a flipping coin.

It doesn't take too much stock market volatility to give consumers the jitters. In an uncertain market, discouraged consumers—especially poorer consumers—will put off spending. In America, consumption spending makes up almost two-thirds of the economy, so when the market takes one of its rare sharp turns, perhaps spooked by discouraging events, such as a near collapse of one of the most respected banks in the world, it's possible to skid off track. Without the guardrails of the Federal Reserve or other interventions of the Treasury, the entire economy could jettison itself into a grave slump, an acute recession, or a mild depression. With recession comes unemployment and with unemployment comes decreased consumer spending and with decreased consumer spending comes. . . . It's not easy, or possibly not even possible, to forecast a depression.[8] We would not be able to predict a depression now any more than we could have predicted the Great Depression, and, according to Simon Johnson at the MIT Sloan School, there was no apparent reason to foresee the scale of the 2008 crisis.[9] Everyone seems to cling to the conservative adage that in the end markets will correct themselves. Possibly, but who can predict how far it is to the end? The expected heads and tails distribution in a long sequence of coin flips will also almost certainly correct itself—but when?

What was needed was a refinancing of the U.S. banking system, a huge injection of capital into the banking system by public money, which, of course, had to be borrowed. Will that work? Ha! That's another gamble. Treasury Secretary Henry Paulson gambled, too, when he permitted Lehman Brothers to fold. He and others pondered the choices of a financial bailout, with all its predictable condemnations, against a more conservative choice of simply letting the giant financial institution sink on its own and letting some gamblers win by cashing in their credit-default swaps.

In the spring of 2008 I had a telephone interview with Damon Rein, a New York hedge fund trader. The United States was in the middle of its sub-prime mortgage mess and the Bush administration was doggedly denying all the obvious indications of the impending recession, claiming that the economy was fundamentally sound. That was before the global financial crisis of September. It was then that I learned just how close Wall Street was to Vegas.

"Hedge funds are just cheap bets on credit," Damon admitted. "Every day, I put money—millions of dollars—on futures and credit-default swaps. It's a legal crap shoot, but you gotta know what you're doing."

"By hedging do you mean buying short?" I asked.

"You mean selling short. I deal in CDSs [credit-default swaps] on bonds. I don't sell short. Well, that's not exactly true. I sometimes short sell to hedge a security that I think is underpriced relative to its fair value. But mostly I deal in CDSs."

He confessed that credit-default swaps are simply wagers. You think Company X will default on its bond payments sometime within the next five years, so you buy a CDS on Company X worth, say, $1 million for an annual premium payment of, say, $100,000. If the company defaults in the fifth year, you've earned $500,000. If it doesn't, well, you've lost $500,000. It's more complex than that, but simply put, you've *swapped* the risk on your own bonds with the company from which you purchased your CDS. If you're worried that the issuer of some bond you're holding will not be able to pay—that it will default on its credit—then you buy CDS to cover your credit risk.

"I thought that they were hedges against losses to the bondholders. You mean you don't have to own the bonds from Company X? Because if you did, and the company defaulted, you would have to subtract the loss of the bond payment from your winnings."

"No! That's the beauty of this. All you need is a hunch that Company X will default!"

"A hunch?" I asked, stunned.

"Yea, you develop an instinct—"

I was surprised to learn that credit-default swaps are in fact wagers, not insurance contracts. I thought that the hedge was a kind of insurance policy on risky speculation. But you don't have to own any bonds that apply to the CDS. I later learned that some CDSs were credit risks in themselves. In 2007, the hedge fund trader John Paulson (one of the Forbes 400 wealthiest Americans and no relation to Treasury Secretary Henry Paulson) used CDSs to wager that sub-prime mortgage bonds would default. They did, and he made a whopping net profit of $15 billion. For Paulson, CDSs were simply gaming contracts. When Lehman Brothers collapsed, so did the payoffs on CDSs purchased from Lehman Brothers. If you bought a CDS contract from Lehman Brothers against Washington Mutual, for example, you should have collected a bundle at the time Washington Mutual closed shop; however, by that time Lehman Brothers was no longer in business to honor its contracts.

Unlike short selling, where you're betting that values will fall, you're betting that the company will default, as long as it doesn't encourage risky behavior by causing the bettor to feel too comfortably insured. Some bettors insure with companies that insured themselves with other companies. Disintegration of the biggest Russian doll exposes the next smaller one.

AIG was one of those dolls. It was designed to survive any conceivable economic calamity, including revolution or a nuclear attack. Yet no one foresaw the debacle that was about to happen.[10] AIG started speculating in CDSs tied to mortgages, gambling with risky complex financial products. After all, all the risk models seemed to show the same thing: every scenario—based on the mortgage history of American homeowners—showed that the vulnerability was small. But with prices soaring and wages stagnant, millions of homeowners could no longer afford to pay their mortgages.[11] The scary thing was that the value of outstanding CDSs in the marketplace had been growing exponentially since 2001, nearly doubling every year. In fact, that value had already exceeded both the world gross domestic product as well as the value of all stocks on the New York Stock Exchange (NYSE).[12]

It comes down to this: Suppose there are three companies, say, *A*, *B*, and *C*. First *B* buys from *A* a credit-default swap on *C*; in other words, *B* places a bet with *A* that *C* will default on its payment of bonds. Let's also suppose that, because *C* has an excellent Moody's rating, *A* believes that *C* is solid and therefore is willing to sell a credit-default swap to *B*. (By the way, this is done without any regulation or oversight.) The benefit? Company *A* brings in free money through premiums that keep the contract going—lots of money. Company *B* may even own some of the bonds issued by *C*. In effect, it is *swapping* the default risk on *C* with *A*. Now here's the problem: Company *C* may do the same, possibly selling CDSs on *B*, or even selling on *A*. Add other players to the game and the whole finance industry becomes a house of cards, where one company could bring down all those that have been wagering on the defaults of others. Just imagine the wagering activity on CDSs over General Motors and Chrysler when they both teetered on the edge of bankruptcy. Most CDS holders are betting on someone else's debt. Nicholas Varchaver and Katie Benner, of *Fortune*, illuminate the predicament.

> And as long as someone is willing to take the other side of the proposition, a CDS can cover just about anything, making it the Wall Street equivalent of those notorious Lloyds of London policies covering Liberace's hands and other esoterica. It has even become possible to purchase a CDS that would pay out if the U.S. government defaults. (Trust us when we say that if the government goes under, trying to collect will be the least of your worries.)[13]

One other concern is that the house of cards is worldwide. There are millions of unregulated CDS transactions involving trillions of dollars stretching through a vast, tangled network of banks and brokerage houses all around the planet. Referring back to our companies *A*, *B*, and *C*, we add Company *D*. Suppose *D* wants a seat at the game. However, also suppose it's a bit late and word on the street is that the betting odds on *B* defaulting have increased. So the price of CDSs on *B* must increase. *A* now has the incentive to take a more reckless risk and sell more CDSs on *B*. Plus, there may be other firms willing to sit in on the game of selling CDSs on *B*. And there we have another house of cards.

CDSs were introduced in 1996 to manage speculation by sharing risk between institutions. Until 2001 the size of the market was relatively small—just tens of billions. Here's another problem: It is true that most economic slumps are caused by complex multiple factors, and the recession of 2008 was no different. However, in the case of that recession, it seems clear that CDS gambling by hedge funds, banks, insurance firms, and other financial enterprises was central to the crisis.[14] In the 1990s banks would buy and sell mortgage-backed securities. But by 2006, with the rise of mortgage defaults, mortgage-backed securities were difficult to sell. And so banks were forced to hold onto them. Nervous about the omens of increasing mortgage defaults, they hedged their risk by buying CDSs from other banks that also owned mortgage-backed securities. Each bank tried to protect itself by sharing the risk with other banks that, in turn, already had their own problems with sizable exposure to mortgage default.

Also, the old market theories were no longer working. Markets were more complex, more non-linear, and more chaotic than theorists had assumed; a response in one part of the market affects a response somewhere else. The old theories would have assumed that the market acts rationally, expelling the gamblers who would be losing somewhere along the way. The central theory implicitly assumed that the markets tended toward equilibrium and that perturbations were mostly random with occasional minor shake-ups caused by unexpected external events. New economic theories were needed to take into account the emotional aspects of change—assurance that one will continue to have a job, faith in banks, confidence to invest, trust in the future.[15]

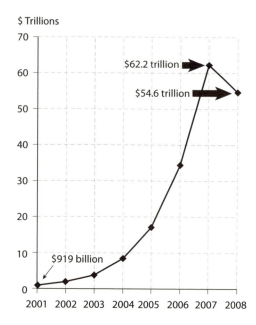

FIGURE 5.1. The value of outstanding credit default swaps, September 2008. *Source*: Mid-Year Market Survey report issued by ISDA (International Swaps & Derivatives Association, Inc.), September 24, 2008. See http://www.isda .org/press/press092508.html.

Another problem was the hedge funds themselves. Some were stable, secure banks and brokerage houses, but others were simply startup companies that were insuring risks without having sufficient backup equity. In a sudden explosion of defaults, all they could do was close shop and go bankrupt, and that's exactly what many startup hedge funds did.[16]

It all happened so quickly. In September 2008 there was a system-wide crisis. By October, the world was about to go through a financial crisis more severe and unpredictable than any it had seen since the Great Depression.[17]

We have to ask the question, *why 2008?* Figure 5.1 tells part of the story. In 2001 the value of outstanding CDSs was *only* $919 billion. Under the Bush administration, outstanding values doubled every year until 2007, when the sub-prime mess hit. In just one year $7.6 trillion was either called or lost as banks, insurance companies, and brokerage houses began to default. Just to give some indication of what $62.2 trillion is, consider that the estimated value of the GDP of the whole world in 2008 was $54.3 trillion, the value of all stocks traded on the NYSE in 2008 was approximately $25 trillion, and the U.S. national debt as of October 2008 was a mere $10.5 trillion.

Two critical questions emerge: "Who was minding the store when CDS trading began growing exponentially?" and "Where did that $7.6 trillion go?" The answer to the first is the Bush administration along with Congress, which did not see the problems coming from this sort of gambling and ignored its responsibility for providing legislation insisting on CDS regulations. In fact, Congress passed a Republican-sponsored bill prohibiting all federal and state regulation of CDSs. The answer to the second is that, though one might argue that $7.6 trillion went back into the whole economy, it benefited only the rich who could afford to buy CDSs through their unregulated hedge funds. In the aftermath, the small investor saw his or her account wiped clean; for in the end, losses trickle down quicker than gains. The sad part of the story is that this game moves money from one side of the equation to the other without productive gain. It built no new planes, new cars, or new washing machines, made no improvements to search engines, and did not contribute to solving any problems that matter most—the environment, world hunger, and peace. It produced nothing but destruction of what was, for all practical purposes, a healthy economy. And we can blame much of it on gambling under the influence of reckless greed—the deliberate courting of danger. Whatever the blame, it's not the math.

Part II
THE MATHEMATICS
♠ ♥ ♣ ♦

Gambling is not done in theory but in the real world where honest dice are carefully manufactured for fairness and where roulette wheels and balls are machined to supreme tolerances. Mathematical models give us that critical, mysterious connection between the ideal theory and the indeterminate physical world. The truly amazing thing is that we can make that connection. Bridges stay up and dams hold because they were built through reliance on mathematical models. Chance seems to be far more mystifying and immeasurable, far more connected to some unfathomable uncertainty; yet, we have truly amazing theorems dealing with likelihood, which tell us that we can cleverly measure nature's secrets of randomness, confine uncertainty by mathematical means, and manage the phenomenology of chance by probability theorems such as the law of large numbers. Who would have dreamed that that relative latecomer to the mathematics world, probability theory—a field born from gambling questions—would turn out to be so essential for describing, using, and controlling so much of nature? Indeed, used properly, it can even be used to value risk and manage luck. But that is no longer a great surprise.

♣ CHAPTER 6 ♣

Who's Got a Royal Flush?

One Deal as Likely as Another

At Monte Carlo is the most sacred shrine of the
goddess; in the directors and croupiers of the famous
establishment are to be found her high priests. There, if
nowhere else on earth, Chance reigns supreme.
—*Karl Pearson, The Chances of Death*

Real-life statistical experimentation vis-à-vis the mathematics of
probability is a beauty that God must have thoroughly enjoyed if
he ever did play dice with the universe. If you have not seen this glo-
rious connection between statistics and probability, or if you missed
it—as I did—trying to learn it in school, then look out. You are in for
one of the great intellectual surprises of the world.

Long ago, in the middle of the last century, I learned probability
and statistics very badly. My professor was a gentle man, the only
person I had ever known with a truly photographic memory. When
he lectured he would peer into the space beyond his delicate frame-
less spectacles and lecture with his head motionless and eyes moving ever so slightly from side to side down a page of the ghost of a
book in front of him. When he came to the end of his page, his right
index finger would tweak as if it were turning the imaginary page.
My admiration for the man extended only as far as his unique gift,

which tempted me with jealousy. This kind, likable young man, in a suit too large for his spindly body, melodically delivered his "how to" lectures—hypnotic catalogues of formulas, methods, and lists of *hows*. Students loved his class; they couldn't be bothered with why things worked when what they wanted was how. To me, the course was a never-ending bore, and so too, I thought, was the mathematics of probability and statistics. I was very wrong!

That was forty years ago. But on one warm autumn morning the next year, I walked north along Broadway, passing newspaper hawkers stacking papers at corners of busy intersections, getting ready for the morning rush. I entered the Columbia University campus, walked into a probability class, and took a seat on a fixed wooden bench behind a long narrow table in a windowless room to hear a lecture being given by a professor whom, because of his remarkable resemblance to that historical figure, students called Lenin. I sat through half the lecture understanding little, amazed at his highly animated enthusiasm for a subject I thought was dull.[1] With passionate blazes of excitement the professor would jitterbug with arms flailing and eyes widening while his head slowly panned across front rows stopping only for brief stares at select students.

"At the start of every physical event," Lenin declared, "chance will delicately direct the event's final destiny and even the most sensitive perturbation of the start may radically alter its final fate. If you hit a billiard ball, extremely sensitive deviations of the angle of the cue stick can cause drastically different results."

I recall the lecture and can even picture Lenin whirling his arms in front of a blackboard talking about billiard cue balls, and Go stones, trying to introduce probability and statistics to a group of students who barely knew the difference between a mean and a mode. With his arms and hands positioning the ghost of a cue stick he talked about billiard balls colliding with object balls,

"The cue ball may hit the 2 ball which, in turn, may hit the 4 ball into the right pocket," he said while his hands worked an imaginary cue stick. "Yet, with the slightest perturbation, the 2 ball may completely miss the 4 ball. To precisely predetermine the initial angle of the cue stick, one must battle all the chances of the minutest conditions that may upset the hit. This is what sports are all about. Every time a rack of pool balls is broken, there is a new game, a game that has never, ever, been played before."

Lenin designed his classes around physical experiments. One day, a large glass bowl of dice was passed around the classroom. Students were instructed to pick one die from the bowl, to toss it sixty times, and to record the number of times it fell on each side.

"Toss your die," he ordered. "But before you do, let me tell you that one of you has a die that is biased toward the three. Which one of you has the one biased toward the three?"

An ideal die should predictably fall every time without bias, but a real die is subject not only to imperfections of manufacture but also to sensitive conditions of the throw. Did it roll along one axis as it left the palm of the hand? How did it hit the surface—rolling along the 2, 3, 4, and 5 faces? If the answer is yes, then a 2, a 3, or a 4 is more likely to turn up than a 1 or a 2.

No matter how fairly a die is made, we should not expect it to be in exact agreement with a mathematical die. If you rolled a real die sixty times and three turned up half the time, you would be suspicious that the die was biased toward the three.

In that class we learned that the model of the idealized die could be used to test the fairness of the real one. In other words, we should presume that the practice of the real world behaves as the idealized theory of the mathematical world so that a real fair die should behave very much like an idealized mathematical die. We learned that we should not ignore how the die is thrown. Perhaps the thrower's thoughts *could* influence the outcome and that any test of fairness must be a test of comparison between the real die and the mathematical one.

Some of us thought that if we tossed our die sixty times and each of the six numbers came up *nearly* ten times then that die would have been fair. But no one knew what *nearly* meant in such a test. If nine 2s came up and eleven 3s came up, would that be a good enough *nearly*? What if eight 3s came up? Surely some deviation from the expected number should be expected. Most of our numbers were strangely close to ten 3s, just what the idealized model predicted, but some rounds wildly differed from ten threes. Others fell far from the predicted outcome, yet close enough to make it difficult to decide their fairness.

Until then, I thought mathematical questions had definite answers, if they had any answers at all. *Nearly* was not a word in my mathematics vocabulary. I expected it to mean something extremely imprecise. To my great surprise, it didn't. By the end of the class we learned that *nearly* meant that after sixty tosses, the number of twos would have to be greater than fourteen to conclude with some probability of being correct that the die was biased toward two. That seemed like a generously relaxed *nearly*, but I had recently read an essay by Sir Ronald Fisher called "Mathematics of a Lady Tasting Tea" in James Newman's *The World of Mathematics*[2] that told the story of an English lady at a tea party who claimed that she could tell by taste whether milk had been added to her cup before the tea or after. No doubt, that would take a finely discriminating palate. It would be harshly inhospitable to hold her precisely at her word, so her claim was graciously interpreted as a more relaxed statement, suggesting that she should be granted some slip, but that most often she could distinguish whether milk had been added before or after the tea.

So an experiment was performed with eight cups of tea, four with milk added before the tea and four after. Clearly, if she were right with all eight cups, the experimenters would be convinced that she could discriminate. But what if she missed one? Would that contradict her word? Maybe not, but what if she missed two?

Nothing is 100 percent certain in this real world of atoms and molecules. Therefore, we must have a way of determining not what's certain but what is probable. The lady should have permitted herself some possibility of error. After all, her taste buds would have changed after the first few sips; the milk would have changed, too, waiting for her tasting. With such a delicate difference between tea poured before milk and tea poured after milk, it seems only fair to relax the notion of certainty and permit a few errors. Fisher's essay is really meant to be about the design of experiments and the concern over subjective error, but here the story is used to point to the connection between mathematics and experiment.

Theoretical mathematics of an idealized pair of dice could predict the behavior of real dice thrown by a real person. Sure, the dice are imperfect white cubes with rounded edges, doubtlessly made in such a way that the indented black dots do not disturb their rotational symmetries. The typical board game die has its pips gouged from the sides of a cube. Each pip is as deep as the next, so the side with six pips is lighter than the side with one pip. Such a die is dishonest, as it favors heavier sides. To make an honest die, material gouged from one side should weigh the same as the material gouged from any other side. The paint to make the pips should also be weighed and balanced.

Let's look at it this way. In geometry we think of points and lines as if they have no thickness; a point is an abstract location. Even a line drawn by the sharpest pencil in the world, scored along the straightest straightedge in the world, is a clumsy human representation of the mathematical line. And yet geometry seems to work well as a model for many real-world applications when measurements are precise enough. Idealized mathematical models may be useful even when they just *nearly* fit the real-world phenomena they are designed to explain. But how near do they have to be? The probability of rolling snake eyes with a pair of dice is 1/36, but don't always expect to roll snake eyes once in 36 tries, even with an ideal pair.

Six blackboards covered the front wall. Obscure mathematical fragments half covered the dusty boards with a dense collage of symbols and diagrams. Lenin had prepared another glass bowl filled with black and white Go stones from the Japanese game of Go.

"There are a thousand Go stones in this bowl," he announced. "How many are black?"

The bowl was passed around. Students were told to mix the stones, pick twenty without looking, call out the number of black stones picked, and then return the stones to the bowl. Lenin drew a horizontal line on the blackboard with equally spaced marks labeled 1 through 20. As the bowl was passed from one student to the next, he would draw a square box over the number equaling the number of black stones called out.

"You should expect each handful of stones to reflect what's in the bowl," he said, "provided that the number of stones in the hand is large enough. But what handful is large enough?"

He meant that a handful of twenty stones may show six black ones, if the true (but unknown) ratio of white to black is close to 14 to 6, but that a handful of forty should be more likely to show twelve black and a larger handful should yield a more precise reflection of the actual number of blacks in the bowl. His point: As the handful number increases it must approach the total number of stones in the bowl; after all, if the handful were so large as to scoop up all one thousand stones in the bowl, it would surely contain exactly the number of black stones in the bowl. No black stones in a twenty-stone handful would be surprising, but *no* black stones in a handful of ninety would be astonishing. In this case, a handful of ninety should have contained a number of black stones in the neighborhood of twenty-seven.

"Isn't it amazing!" shouted the professor, once again whirling his hands in the air. "The real world seems to like to stick wonderfully close to mathematics!"

A few days after attending that class, quite by chance, I ran into Lenin at Le Figaro, a café in Greenwich Village. It was a smoke-filled place with walls covered with collages of yellowing editions of the French newspaper *Le Figaro*, mirrors, framed photos of local poets, and antique menus showing what a nickel could buy in the 1920s. We talked.

I knew that the odds of drawing a royal flush of, say, clubs are 2,598,959 to 1 but wondered why it was any different from drawing any poker hand. Why such a likelihood? There are 52 distinct ways of picking the first card, 51 distinct ways of picking the next, and so forth. Therefore there would be $52 \times 51 \times 50 \times 49 \times 48$ distinct ways of picking five cards. However, that does not take into account the order in which those cards were picked. For example, the hand may be arranged as **A♣ K♣ Q♣ J♣ 10♣** but be picked in any order. The ace of clubs may have been dealt first, second, third, fourth, or fifth. Fixing when the ace is dealt leaves four possibilities for the king, three for the queen, two for the jack, and one for the ten. Hence, in order to compute the number of ways that a royal flush can be dealt, we must divide $52 \times 51 \times 50 \times 49 \times 48$ by $5 \times 4 \times 3 \times 2 \times 1$. That computation turns out to be 2,598,960.

Surely, I thought, any one hand is as likely as any other, but it didn't seem right. Four different royal flushes are possible, one for each suit. So the odds of getting any one of those royal flushes are 2,598,956 to 4, but the odds of getting a particular one—say, a royal flush of diamonds—are still 2,598,959 to 1. Clearly it's harder to get a royal flush of spades (A♠ K♠ Q♠ J♠ 10♠) than it is to get an unspecified royal flush of any suit. Any specific dull hand is also rare and there are loads of dull hands. For example, anyone would agree that the hand 3♠ 6♥ 8♣ J♦ Q♠ is dull. But the odds of being dealt that precise hand are 2,598,959 to 1. So, paradoxically, it should be amazing to get a dull hand.

I learned that I had been thinking wrongly. Suppose ten people are in a poker game. Then the likelihood of a royal flush being dealt to any one player is much higher than the likelihood it will be dealt to any one particular preselected player. In other words, that same royal flush may have 2,598,959-to-1 odds when viewed as simply being dealt to any one of the ten people playing, but much smaller odds when we specify who should get it.

There are only 4 possible royal flushes, 36 straight flushes (five descending numbers of the same suit), and 624 hands that are four-of-a-kind. It may be that the likelihood of getting a particular royal flush is the same as that of getting a particular straight flush, or of getting a particular four-of-a-kind, or of getting any other particular dull hand, but the likelihood of getting any one of the four royal flushes is surely smaller than getting any of the 36 straight flushes. It's even less likely to get any of the 36 straight flushes than to get any of the 624 four-of-a-kinds. One deal is as likely as any other!

Lenin gave me the clue to understanding.

"Experiment with a deck of cards," he advised. "Mark a black dot on each of the cards you wish were part of your hand. If you want to draw a royal flush, put a black mark on the five cards ace, king, queen, jack, ten—say, all diamonds. Then drawing a royal flush is like picking five black Go chips from a bucket of 52 chips, 47 white and 5 black. It's simply that the probability of drawing the 5 black chips is the same as drawing a royal flush."

I understood that the probability of drawing the first black marked card is 5 out of 52 because I would be drawing any one of the 5 marked cards. But on the next draw, the probability of drawing another black marked card is 4 out of 51 because I would have removed one card from the deck. And on the next draw it would be 3 out of 50. In the end, the probability of my drawing exactly 5 black marked cards is

$$\left(\frac{5}{52}\right)\left(\frac{4}{51}\right)\left(\frac{3}{50}\right)\left(\frac{2}{49}\right)\left(\frac{1}{48}\right),$$

which works out to be

$$\frac{1}{2,598,960},$$

the probability of my drawing a particular royal flush—say, a royal flush of diamonds. Here we are using the fact that the probability of two independent events happening is the product of the probabilities of the separate events. So the probability of drawing the first marked card is 5/52. The probability of drawing the next marked card is 4/51. So drawing both marked cards is the product

$$\left(\frac{5}{52}\right)\left(\frac{4}{51}\right).$$

Continuing this way we find that the probability of getting all five marked cards is

$$\left(\frac{5}{52}\right)\left(\frac{4}{51}\right)\left(\frac{3}{50}\right)\left(\frac{2}{49}\right)\left(\frac{1}{48}\right).$$

It is the same as the odds of putting black dots on exactly the five cards that make any particular one of the four royal flushes. The odds of drawing a royal flush of diamonds are 2,598,959 to 1. The chance of drawing any of the four possible royal flushes is four times better, 2,598,956 to 4 or, more simply, 649,739 to 1.

CHAPTER 7

The Behavior of a Coin

Making Predictions with Probability

I'm gamblin',
Gamble all over town. . . .
Yea, where I meet with a deck of cards,
Boy you know I lay my money down.
—*Lightnin' Hopkins, "The Roving Gambler"*

My great books course professor relayed an anecdote about Flaubert. The professor himself wore a stained, tan trench coat during the hour and, with his difficulty of getting Bs out without stuttering them, told our class that Flaubert, *the author of Madame B-B-B-Bovary*, had a catalogue of writing exercises that he shared with his friend Maxime du Camp.[1] The two would have lunch in small Paris restaurants and often would try to match the coats on the coat rack with the clientele of the place, just by observing features, mannerisms, and expressions. I have no idea how true or apocryphal the story is, but anyone reading passages of *Madame Bovary* would well believe it. See how carefully Flaubert describes Charles Bovary, a young boy about to enter lycée.

The Newcomer, who was hanging back in the corner so that the door half hid him from view, was a country lad of about fifteen,

taller than any of us. He had his hair cut in bangs like a cantor in
a village church, and he had a gentle timid look. He wasn't broad
in the shoulders, but his green jacket with its black buttons seemed
tight under the arms; and through the vents of his cuffs we could
see red wrists that were clearly unaccustomed to being covered.
His yellowish breeches were hiked up by his suspenders, and from
them emerged a pair of blue-stockinged legs. He wore heavy shoes,
hobnailed and badly shined.[2]

In this description, Flaubert is setting us up with a heuristic representation of a boy who will very quickly become a country doctor
in the novel. In fact, we are not surprised to learn that this country
doctor is merely an *officier de santé*, a person schooled in medicine not
holding a medical degree.[3]

You are wondering why I bring this story up in a book on gambling. Flaubert's novels are filled with heuristic representations that
make us believe we are right when we match the coat with the person.
It is what all good novelists do. Reading novels is often an exercise in
making heuristic judgments. In real life we make such matches all
the time. They are the routine judgments we make, and they come
from beliefs driven by stereotyped circumstances. Without conscious
awareness, we make probabilistic judgments about people, events,
and outcomes based on a very limited number of vague and dubi-
ous assumptions coming from our internal hypotheses. Suppose you
read that a fellow by the name of Steve is very shy and withdrawn,
invariably helpful, but that he has little interest in people or in the
world of reality. He is submissive with a need for order and structure.[4]
Now order Steve's most likely occupations from the following list:
farmer, salesman, airline pilot, librarian, and physician. We have a
strong sense that if the list includes Steve's true occupation, then he
is most likely to be a librarian and least likely to be an airline pilot.
Why shouldn't Steve be an airline pilot? Or, more to the point, why
shouldn't the job of pilot be the most likely—consider his need for
order and structure, as well as his passion for detail.

Steve seems to have been represented as the stereotypic librarian, not as the stereotypic airline pilot. The problem here is that two

measures are confused—likelihood and stereotype. Research shows that most people tend to have that confusion.[5] But there is something else going on here. Shouldn't it be more likely that Steve is a farmer? After all, there are many more farmers than librarians. If we simply evaluate Steve as a man obeying some random frequency distribution, he would more likely be a farmer. However, under normal heuristic tendencies the probabilistic frequency of farmers in the overall population does not trump considerations of stereotypic similarities, and so we think that he is more likely to be a librarian. In 1973, Amos Tversky and Daniel Kahneman conducted research to study how subjects consider probabilistic as opposed to stereotypic judgments. There were two groups. In one, the subjects were given character descriptions and told that they were from a pool of 70 engineers and 30 lawyers. The other was given the same character descriptions and told that they were from a pool of 30 engineers and 70 lawyers. The subjects in the first group would have a better (probabilistic) chance of being right by simply guessing the character to be an engineer. Similarly, the subjects of the second group might have guessed the character to be a lawyer. But the result of the study determined that people tend to ignore the odds and bank on the descriptive representation of character with no consideration of majority.

More surprisingly, majority representation is ignored, even when the description is uninformative. The following description contains no information about the character's occupation: "Dick is a thirty-year-old man. He is married with no children. A man of high ability and high motivation, he promises to be quite successful in his field. He is well liked by his colleagues."

We should expect that subjects would consider the odds of 7 to 3 in favor of being an engineer if the pool contains 70 engineers and 30 lawyers. But that's not what happens. Subjects simply gave Dick even odds of being an engineer or lawyer.[6]

This discounting of relative pool sizes reflects an insensitivity to sample size. Consider the answers to the following question by 95 undergraduates after they were told that 50 percent of all babies are boys and that in a certain town there is a large hospital where about 45 babies are born each day and a small hospital where about 15 babies are born each day. The students were told that for one year each hospital recorded the number of days in which more than 60 percent of the babies born were boys and asked the following question:

> Which hospital do you think recorded more such days?
> - The larger hospital (21)
> - The smaller hospital (21)
> - About the same (that is, within 5 percent of each other) (53).[7]

The surprise here is that students are not paying close attention to the probabilistic logic that suggests the smaller hospital is more likely to report more days when the male proportion exceeds 60 percent. This is simply because the larger hospital has a larger sample and is therefore less likely to deviate from the norm.

Such studies point to misconceptions of chance. People tend to misinterpret the law of large numbers as saying that a long sequence of reds in the spin of a roulette wheel should favor black's turn. You might think that the sequence R-B-R-B-B-R is more likely than the sequence R-R-R-B-B-B, simply because this latter sequence does not appear to suggest true randomness. According to Tversky and Kahneman, people expect that the important characteristics will be represented in each of its specific local parts.[8] Yet, that's not how chance works! Representations of smaller samples deviate from the global expectation. We expect local and regular self-corrections, whereas what we find is simply a gradual dilution of local events as the number of trials increase.

One further point to be raised is the notion of anchoring. We all have a tendency to anchor our opinions to some immediate bias. Consider another experiment conducted by Tversky and Kahneman in which subjects were asked to guess the percentage of African countries listed as members of the United Nations (in 1974). A wheel marked with numbers from 0 to 100 was spun. The wheel would come to rest at a number, say X. The subjects were asked to first indicate whether X was higher or lower than the answer to the question. Following that, the subjects were asked to estimate the value of the quantity by moving upward or downward from that number. The bizarre outcome was that, for the group that saw the wheel land on 10, the median estimate of the percentage of African countries that were members of the United Nations was 25, and for the group that saw the wheel land on 65, the median estimate was 45. Now what does a wheel of fortune have to do with the number of countries belonging to the United Nations?

It seems that humans have definite cognitive biases that are affected by how questions are framed. We might expect that these biases influence a layman's decisions; surprisingly, they also influence the intuitive judgments of the experienced researcher as well.[9]

Our more rational judgment embraces underlying probabilities and hence a more rational understanding that perhaps there are some mathematical models that could be used to enhance our judgment of the stochastic world. When we consider flipping a fair coin we all know that the probability of it coming up heads is 1/2. The law of large numbers tells us that the ratio of the number of heads to the number of tails approaches 1 as the number of flips grows larger. Heuristic judgment muddles the meaning into a belief that somehow a long string of tails will be made up by a balancing string of heads. The general public continues to confuse the proper meaning of the law of large numbers with the feeling—and erroneous belief—that if a face has not come up for a very long time, the chances of its appearance increase with every turn. And yet that same public knows that theoretically, every time a coin is flipped and every time a roulette wheel is spun the odds against each outcome are precisely the same—the coin is just as likely to land on heads as tails and the roulette ball is just as likely to fall into any one stall as into any other. It's just that people tend to muddle the difference between outcomes and frequencies.

Toyota Prius drivers have played with this law. The Prius displays the average miles per gallon and that figure can be reset. When starting out on a journey of, say, 500 miles, the displayed number fluctuates depending on driving conditions. In the first 10 miles the display may read, say, 52.3 miles per gallon. At 20 miles it may read 46.4 miles per gallon. New Prius owners tend to *hypermile*; that is, they tend to drive in a manner that maximizes the average miles per gallon. Generally this means managing the gas pedal with an awareness of its effect on the miles per gallon on the display screen. They quickly notice that the fluctuation is wild for the first ten, twenty, thirty miles, and that that wildness tends to dampen into a stable range that narrows more and more toward some limiting number. By 200 miles, the display may read 51.2 miles per gallon. At that point even the most skillful hypermiling will not change the display by more than a few tenths. This is because the car's miles per gallon history is on the side of the accumulated average. If the average of ten numbers is 46.4 and we throw in five more numbers, say, 59, 56, 62, 58, and 57, the average of the fifteen numbers is 50.4, a difference of 5. However, if the average of one hundred numbers is 46.4 and we throw in those same five numbers, the average is just 47, a difference of 0.6.

Figure 7.1 represents the cumulative outcome of 10,000 repeated coin flips (+1 for each head and –1 for each tail).[10] The horizontal line represents 0. So, for example, after approximately 2,500 flips, the outcome is approximately –20. In this particular run, tails is in the lead almost 97 percent of the time. It is a horse race between two horses of equal odds. Normal intuitive judgment favors the opinion that the graph should bounce over and below the zero line far more often than is pictured. However, true mathematical theory tells us that it is far more likely for the graph to favor one side over the other for relatively long periods of time.[11] The reason is that once the cumulative outcome strays far from 0 in the negative direction, it needs long strings of heads to get it back to positive territory.[12]

The cumulative record of coin flips may be modeled through a Galton board. Sir Francis Galton, the nineteenth-century English geneticist, constructed a board filled with pegs arranged with a funnel at the top and chambers at the bottom, as in figure 7.2. Galton's point was to demonstrate that physical events ride on the tailwinds of chance. The ball falls to the first pin and must decide whether to fall left or right. Being unstable in its decision, it does something equivalent to flipping an unbiased coin: heads go left, tails go right. Whichever way it goes, it falls to the next pin and must decide all over again. What does this really mean? Imagine the perfect Galton board as one in which the balls always fall directly on the absolute tops of pegs. What makes the ball fall to the right or left? We said

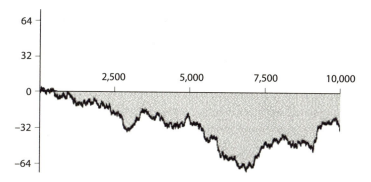

FIGURE 7.1. Cumulative outcome of 10,000 repeated coin flips.
Source: "Leads in Coin Tossing," http://demonstrations.wolfram.com/Leads
InCoinTossing/. With permission under the guidelines of http://creative
commons.org/licenses/by-nc-sa/3.0/.

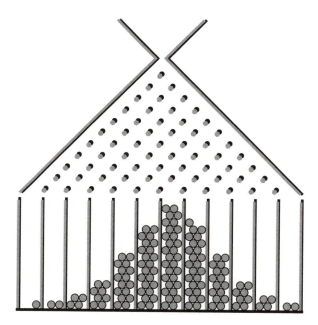

FIGURE 7.2. A Galton board.

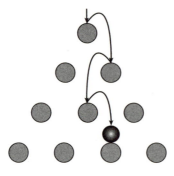

FIGURE 7.3. A bounce off the top peg of a
Galton board.

that it was like a coin flip. But that flip may be determined by a history of dependent events initiated by a minuscule perturbation, such as a butterfly flapping its wings over the Pacific or a cow farting in an Idaho cornfield. Before each coin flip, the outcome of the previous flip is history; the coin no longer remembers the outcome and therefore will behave as if it is a new coin. However, the cumulative outcome does take into account the history of all previous coin flips.

A bounce off the top peg goes left because . . . well, we don't exactly know why—just because something remotely connected happened, an undetectable air current or vibration. But it represents one of the many unaccounted measurements that imperceptibly affect the outcome. It rolls off the peg to fall directly over the next peg down and, this time, goes right because . . . who knows?

The 50-50 chance of going left or right causes the build-up of the bell-shaped curve marked by the highest balls in figure 7.2. Counting the number of ways the balls can fall proves this. Suppose that a ball is dropped and we mark its descent by the letters L and R to indicate bouncing to the left or right. We would then have the following possible outcomes:

LLLL
LLLR, LLRL, LRLL, RLLL
LLRR, LRLR, LRRL, RLLR, RLRL, RRLL
LRRR, RLRR, RRLR, RRRL
RRRR

There are more combinations of mixed letters than non-mixed and since there is an equal chance for the ball to go left or right there is a tendency for it to end in the center slot of the Galton board.

Every event in nature has to account for a vast number of indeterminate possibilities. The toss of a die may strongly depend on its initial position in the hand that throws it and more weakly depend on sound waves of a voice in the room. Those are just two external modifiers that guide the die to its resting position. How it strikes the table, the precision of its balance, how it rolls off the hand, and the elasticity of its collision with the table will influence which side faces up when it comes to rest.

Returning to the Galton board, let's arbitrarily count falling to the left as −1 and falling to the right as +1. After bouncing down eleven rows of pegs the ball will end up in one of the twelve pockets at the bottom of the board.

So, for example, the ball at the extreme left will end up with a cumulative value of −11. The final position of each ball represents a distinct cumulative outcome. As figure 7.2 demonstrates, the balls tend to accumulate more toward the center than away from the center. However, though quite a few balls fall in the two center slots, more fall in the ten remaining slots. This may seem surprising, given that there are fewer possible paths leading to the outer slots than to the inner slots.

In figure 7.2, the collection of balls represents the final accumulated values of 140 experiments—31 fell in the 5 slots on the left; 55 fell into the 5 slots on the right; and 54 fell into the two middle slots. It is true that the final position of any one ball does not indicate the history of its journey the way the graph in figure 7.1 does. However, there are two critical things to notice here: (1) the first two rows of pegs limit the outcome; left on the first and right on the second (or vice versa) force the final accumulated value to be less than 11 and greater than −11; and (2) 61 percent of the balls have fallen outside the center two slots. Now it is possible for a ball to start out on the left side and end up on the right, but it is also very likely that any ball that wanders too far to the left will have a decreasing chance of returning to the right. In other words the graph depicted in figure

7.1 is highly representative. The Galton board could be modified to alter the odds and give a slight edge toward losing and therefore model typical casino games. For example, the ball may be forced to roll down the right incline before entering the board. That would put a counterclockwise spin on the ball, so when it hits the first pin it has a tendency to move to the left after impact. Then, the center of gravity of the pile at the bottom would be shifted left—the longer the incline, the larger the shift.

Now let's see how this would work in an idealized situation where you have a fair coin and are betting heads double or nothing against the bank. You have, say, $100. The $100 could become $200 on the first flip, $400 on the second, and so forth. The Galton board 11 pegs high and a single ball would be a useful model for what might happen. Suppose falling to the right indicates heads and a win, falling to the left a loss. Since the coin is assumed fair, there is equal chance of the coin ending on the right side as on the left. However, as we have noticed, it is more likely that the ball will not end up in the middle columns, so the player will more likely either win more than $400 or lose more than $400. There is even a chance that the loss will be as great as $6,400. The expected outcome of the first flip is either $200 or $0, with an equal likelihood, so the mathematically predictable result of the wager, including the amount of the initial stake, is $\frac{1}{2}(\$200 + \$0) = \$100$. With such an expectation, why would you bet? After all, you already have $100 and all you can expect is $100. The irrational intuition tells us that if you are lucky you will walk away with $200 after the first flip, but the rational math tells us that you are just as likely to walk away with nothing and the odds are that you will end up with what you already have. But you know that can't be true because there are only two possibilities, $200 or $0. So how should we interpret what the math is telling us?

Should you risk your $100 or gamble for the $200? It is that word *risk* that is at the heart of all gambling. Economists have long sought a meaningful measure of risk, which should depend on a person's specific financial situation; what is risky for the poor is not so risky for the wealthy. Two people play double or nothing with a stake of $10,000. A player with a net worth of $10 million will be less upset than one with a net worth of $10,000. Risk must be associated with value, not exclusively on price. And that value depends on the individual circumstances of the person calculating the risk. Thus there is good reason to define value as a function of utility—that is, how important or useful is the winning to the player's life.

It may seem a paradox that the expected value of the wager does not change with the number of plays. On the second flip, the expected value is $(\$400 + \$0)/4 = \$100$, on the third $(\$800 + \$0)/8 = \$100$, and so forth (because the probability of two consecutive heads is 1/4, three consecutive heads is 1/8, etc.). After playing n times the expected value is therefore n hundred dollars. So, after 11 flips the gambler should expect a return of $1,100. This may seem absurd; shouldn't the expectation of loss and gain be equal? Ah, but we have not considered the risk of losing.

In 1738 Daniel Bernoulli, nephew of Jacob Bernoulli and pioneer in probability theory, wrote an essay on the subject of cumulative outcomes of coin flipping; there had been a great deal of speculative ideas on the subject by mathematicians before him.[13] In this essay Bernoulli referred to a problem that his cousin Nicolas Bernoulli once submitted to the mathematician Pierre Rémond de Montmort, the following poser.

> Peter tosses a coin and continues to do so until it lands "heads" when it comes to the ground. He agrees to give Paul one ducat if he gets "heads" on the very first throw, two ducats if he gets it on the second, four if on the third, eight if on the fourth, and so on, so that on each additional throw the number of ducats he must pay is doubled. Suppose we seek to determine the value of Paul's expectation.[14]

So the game ends the moment the coin falls on heads. In other words, Peter agrees to pay Paul one ducat if the first throw comes up heads, two ducats if it comes up tails on the first throw and heads on the second; four ducats if it comes up tails on the first two throws and heads on the third; and so forth. And Paul is to pay Peter for the opportunity to play the game.

How much should Paul pay for the privilege of playing this game? Few of us would pay much to enter the game. The game is sure to end sometime, for though, even with very good luck, Peter may throw a long string of tails, the game has a high likelihood of ending in a finite number of throws.

Bernoulli proposed that Paul must consider his marginal utility and risk. He coined the word *utility* to represent usefulness of gain and tried to represent utility by redefining *expectation*, whereby small increases in wealth as a quantity becomes inversely proportionate to the quantity of goods a person already has. For him, utility was how excess money would be utilized to the satisfaction of the gainer: would he eat better, be more comfortable, and so on? His attempt was to level out the value field between rich and poor, so a ducat to a rich person has less utility than a ducat to a poor person. If John is twice as rich as Paul, then John should be half as happy to win a hundred ducats than Paul; in other words, the utility of John's gain should be half that of Paul's. The curious message here is that as Paul continues to play, his gains' utility keeps diminishing because the wealth keeps increasing; on each throw his earnings increase while its utility decreases. Bernoulli was looking for something too precise for the subjective pleasure coming from wealth. Under uncertainty, people act to maximize their expected utility, not their expected value.[15]

Accepting double-or-nothing ventures is risky behavior. However, let's recall that we are talking about a fair coin, where the "odds" of heads are precisely the same as the "odds" of tails. What happens if the odds are slanted in one's favor? Does the omen apply to someone who has clocked the coin and therefore knows something about which side will come up more often? Yes, it does. And that slightly unfair coin is exactly what casinos everywhere bank on.

I am not suggesting that the casinos are dishonest. They don't have to be to make a profit. Recreational gamblers as well as professionals know that their favorite casinos have a small advantage over them, so they bank on that word *small* without considering the thought that the small advantage is also persistent. The advantage is always there. They know that gambling is playing with chance and that chance is always on the side of the casino. The intelligent gambler knows that the indiscriminate enigmatic commands of the goddess Fortune may affect the outcome of an individual event—how the dice fall or how the cards are shuffled—yet also knows that mathematics has something to say about the percentages of successes and failures in the long run. This observation is one of the most remarkable in the history of Western thought.

Those of us who have seen the movie *Casablanca* a hundred times or more will recall the scene where Rick tries to save a young Bulgarian girl's fiancé, Jan, from ruin and at the same time save the pretty, naive Annina from collecting an exit visa from the unscrupulous police captain, Louie Renault.

The scene takes place in the gaming room of Rick's Café. Jan is seated at the roulette table. He has only three chips left. Rick enters and stands behind Jan.

CROUPIER (*to Jan*): Do you wish to place another bet, Sir?

JAN: No, no, I guess not.

RICK (*to Jan*): Have you tried twenty-two tonight? (*Looks at the croupier.*) I said, twenty-two.

Jan looks at Rick, then at the chips in his hand. He pauses, then puts the chips on twenty-two. Rick and the croupier exchange looks. The wheel is spun. Carl is watching.

CROUPIER: Vingt-deux, noir, vingt-deux. (*He pushes a pile of chips onto twenty-two.*)

RICK: Leave it there.

(*Jan hesitates, but leaves the pile. The wheel spins. It stops.*)

CROUPIER: Vingt-deux, noir. (*He pushes another pile of chips toward Jan.*)

RICK (*to Jan*): Cash it in and don't come back.

(*Jan rises to go to the cashier.*)

A CUSTOMER (*to Carl, the bartender*): Say, are you sure this place is honest?

CARL (*excitedly, in his lovable Yiddish accent*): Honest? As honest as the day is long![16]

At the end of the nineteenth century, when the casino at Monte Carlo was still relatively young, a Paris newspaper/journal called *Le Monaco* published records of 16,500 spins of a Monte Carlo roulette wheel during a four-week period in July and August 1892.[17] Before the mathematical statistician Karl Pearson published his analysis of Monte Carlo roulette, everyone believed that the wheel spun according to the expectations and frequencies of probability. Not so! Pearson found that the mechanism, as machine precise and as perfectly adjusted for the table as it could be, was not fully obeying the laws of chance, laws that suggested a tighter frequency around the mode. With complete precision it would be equally likely for the ball to fall in any one of the thirty-seven pockets as any other.

Excluding the 0 pocket, there is an equal mathematical chance for the ball to fall into a red or black pocket.[18] This should mean that in a great many physical spins, the ball should fall into the red pocket about 50 percent of the time. However, Pearson studied a table of 16,019 trials where the ball fell into a red slot 50.27 percent of the time. So, is that 0.27 percent over the expected 50 percent surprising? It represents just 27 more red outcomes than black in 10,000 spins. It seems to be a very small deviation from the expected percentage. It turns out that such a deviation from the mean is likely. No surprise there, so the expected general equality of red and black holds. However, Pearson then turned to the question of the running sequential distribution of reds and blacks. Do the experimental patterns of successive spins conform to theoretical expectations? To explain, we return to Bernoulli's poser, sometimes referred to as the St. Petersburg Paradox.

To investigate the sequence of red and black outcomes on a roulette wheel, Pearson reduced the problem to flipping a coin. If the coin is fair, the chance of it coming up heads is 1/2, very close to the same as the chance of the roulette ball landing in the red pocket. (There are 37 pockets on the wheel including 0. In European roulette there is no 00 pocket—18 are red, 18 black; 0 is green, so the probability of red is 18/37, which is very close to 1/2.)

So the chance of two heads coming up in succession is $1/2 \times 1/2 = 1/4$; the chance of three heads is $1/2 \times 1/2 \times 1/2 = 1/8$, and so forth. Let's define a *run* to be a sequence of heads (or the roulette ball falling into a red pocket). Therefore, the mathematical theory predicts that in n tosses, there would be

$$\frac{n-k+1}{2^k}$$

⑩ runs of length k. For example, in 2,048 tosses of a fair coin, we should expect results described in table 7.1. Paul might have considered this in pricing his wager with Peter.

So, in 2,048 tosses of a fair coin, we should expect just one run of length 11.

That is the theory. In practice, Pearson found the following results after spending a fortnight examining 4,274 spins of a Monte Carlo roulette wheel (see table 7.2). Examining the last two rows we find something strange. For a run of length 1 the actual deviation is almost ten times the size of the standard deviation! The odds against such a thing happening with a fair roulette wheel (as with a fair coin) are more than ten trillion to one! Pearson wrote that if the game were a truly fair game of chance then we should not expect to see such an outcome once in a game that had gone on since the beginning of geological time on this earth.[19]

Okay, perhaps by some chance, Pearson hit on one miraculous fortnight that was so improbable that it could only occur once in the history of the world. Should that be a reason to doubt the fairness of the roulette wheel? His student tried the experiment again for another fortnight and found results not as improbable as Pearson's but ones that would be expected to occur just once in five thousand years of continuous round-the-clock playing. And during another fortnight at Monte Carlo, observing 7,976 spins of the wheel, another investigator, a Mr. De Whalley, computed deviations from the standard deviations giving the odds against a fair wheel at 263,000 to 1. Other experiments found the same miracles. In an 1893 observation of 30,575 spins of a Monte Carlo roulette wheel, observed outcomes

TABLE 7.1
Expected Run Length Using a Fair Coin (2,048 tosses)

Run length	1	2	3	4	5	6	7	8	9	10	11
Tosses	1,024	512	256	128	64	32	16	8	4	2	1

TABLE 7.2
Actual Run Length at the Roulette Wheel (4,274 Spins)

Run length	1	2	3	4	5	6	7	8	9	10	11	12	12+
Experiment	2,462	945	333	220	135	81	43	30	12	7	5	1	0
Theory	2,137	1,068	534	267	133	67	33	17	8	4	2	1	0
Standard deviation	33	28	22	16	12	8	6	4	3	2	1.5	1	—
Actual deviation	325	123	201	47	1	14	10	13	4	3	3	0	0

differed from theoretical outcomes by almost 6 times the standard deviation, odds of more than 50 million to 1. Pearson expressed his findings in this sardonic conclusion.

> Monte Carlo roulette, if judged by returns which are published without apparently being repudiated by the Société, is, if the laws of chance rule, from the standpoint of exact science the most prodigious miracle of the nineteenth century. . . . We appeal to the French Académie des Sciences, to obtain from its secretary, M. Bertrand, . . . a report on the colour runs of the Monte Carlo roulette tables. . . . Should he confirm the conclusion of the present writer that these runs do not obey the scientific theory of chance, then science must reconstruct its theories to suit these inconvenient facts. Or shall men of science, confident in their theories, shut their eyes to the facts, and to save their doctrines from discredit, join the chorus of moralists who demand that the French Government shall remove this gambling establishment from its frontier?[20]

A more recent story involves the big-time Vegas casino owner Steve Wynn and William Walters, the guy who won almost four million dollars in thirty-eight hours of continuous roulette playing at Atlantic City in the summer of 1986. Walters clocked the roulette wheel—that is, recorded the frequencies of winning numbers—and bet on five numbers (7, 10, 20, 27, and 36) at the Golden Nugget.[21] The

Golden Nugget thought that perhaps the wheel was biased and so had it checked by agents of the New Jersey Casino Control Commission and the Division of Gaming Enforcement. No bias was found. Three years later Walters had the wheel at the Claridge Hotel and Casino clocked by his boys. He played that wheel and within eight hours pocketed $200,000. When the casino shut that wheel down he moved to another and won another $300,000. I had trouble verifying this story and so contacted Steve Wynn, who could not or would not say whether it was true or false, but I got the feeling this was not the end of the story and that there were many small casinos with biased wheels. Much can be won through paying meticulous attention to the statistical history of roulette wheels.

But beware the con! Take the story of "Swindled," a young man who contacted me when he heard that I was writing this book. There were many such stories; most were doubtful, but this one passed inspection. Briefly this is what happened. He had just dropped out of Rutgers where he majored in computer science. Failing his exams, he became disenchanted with his studies and drifted for a while between Philadelphia and New York washing dishes and working in bowling alleys. Whenever he had a few dollars in his pocket he would go to Atlantic City and play roulette. One day when he was having lunch in a café in Atlantic City a young girl walked in and sat next to him.

"I want to talk with you," she said, nervously.

For several silent, entrancing moments, he stared at her eyes and thought she couldn't be older than sixteen.

"There is $40,000 in cash here," she said, pulling out a rubber-banded brick of hundred-dollar bills from a large envelope. "I'm Marcia and my mother is dying of lung cancer in Nigeria and I must raise $200,000 for a lung transplant. I would like you to stake this at the Golden Nugget. You get 25 percent of the winnings."

"What if I lose?"

"If you lose, you owe me nothing," she said with a smile. "But you won't."

"Why do you think so?" he asked.

"A friend saw you at the casino last month. She said that you have luck on your side. I also know that you clocked the wheels. There's no risk to you. It's my money." Then, she added, "They won't let me in; I'm underage."

It was true. Swindled had won reasonably big at the Golden Nugget the month before at roulette, but he didn't think of himself as a regular gambler. He did clock one wheel. So Swindled accepted the offer without knowing anything more than Marcia's first name. The very next day he played roulette at the Golden Nugget and, by a scheme of clocking and then playing five selected numbers that seemed to be favored by the wheel, walked away with almost $200,000. As arranged, he met Marcia in a secluded corner of the café and handed her $133,600 in cash.[22]

This all would have been fine. He would have won $31,200 in one day at no risk to himself. But a few months later, after losing almost all his winnings, the FBI arrested him on a counterfeiting charge. He spent three months in jail and $12,000 on lawyer fees to get out.

CHAPTER 8

Someone Has to Win

Betting against Expectation

> The probable is that which for the most part happens.
> —*Aristotle, Rhetoric*

Play a game of chance, any game of chance. It could be flipping a coin, shooting craps, playing roulette, or betting on a horse race. In the end you either lose or win. Let us introduce some notation: *P(A)* will represent the probability that event *A* will turn out successfully. For example, if *W* represents a win and *L* a loss, then *P(W)* is the probability that you win your chosen game and *P(L)* the probability that you lose. If you were flipping a coin, then *P(W)* would equal 1/2 and so would *P(L)*. In American roulette, there are 38 pockets on the wheel including 0 and 00—18 are red, 18 black; 0 and 00 are green. Therefore, if you were betting on red, *P(W)* would equal 18/38 or, more simply, 9/19, and *P(L)* would be 10/19. If you were rolling a die hoping for an ace, *P(W)* would equal 1/6.

Consider this question. If you play the game four times, you could win once, twice, three times or four. What is the probability of winning 0, 1, 2, 3, or 4 times? My question is a fair one, since real gambling involves a cumulative string of wins and losses, often a long string. You would surely want to know the odds of breaking even or better in your bets, for that would be the odds of not losing more than twice in four bets.

To study my question let us agree on the following notation: a string of Ws and Ls will represent a respective string of wins and losses. So losing all four times would be marked by LLLL and winning all four times by WWWW. There is only one way to win all four times and only one way to never win. Note the duality in that last sentence. What about winning once in four rounds? There are four ways that that could happen: WLLL, LWLL, LLWL, and LLLW. And of course by that same duality, we would have four ways of losing once in four rounds. What about winning twice in four rounds? The configurations are now WWLL, WLWL, WLLW, LWWL, LWLW, and LLWW. So, there are six ways of winning twice in four rounds. Notice that in the end, the order of wins and losses does not matter; we listed them in strings of four letters with regard to order simply to count them systematically.

Now let's take the middle case of having two wins in four rounds and ask for that probability. To simplify the notation we shall let p represent $P(W)$ and q represent $P(L)$. The probability of one single win is p, and since wins and losses are independent (i.e., each round does not depend on the round before it), we see that the probability of having two wins in four rounds is p^2q^2. This is because you would have to win twice *and* lose twice, and when the logical connective is *and*, the probabilities are multiplied. But, as we have seen, this can happen in the following six distinct ways: WWLL, WLWL, WLLW, LWWL, LWLW, and LLWW.

So, since the logical connective *or* here corresponds to +, the probability of any one of these events happening is $ppqq + pqpq + pqqp + qppq + qpqp + qqpp$, or simply $6p^2q^2$.

Table 8.1, constructed from knowing the values of p and q for the three different games, shows the probabilities of winning 0, 1, 2, 3, and 4 times in four rounds. In theory, for both roulette and coin flipping, according to table 8.1, a player is most likely to win twice in four rounds. We could construct a table of probabilities for, say, a hundred rounds of roulette and coin flipping, though it would be dauntingly long and impractical. Instead, let me just say that in a

TABLE 8.1
Probabilities of Winning 0, 1, 2, 3, and 4 Times in
Four Rounds of Three Different Games

Number of wins	Number of ways a win can occur	Probability of winning	Probability of red in roulette	Probability of heads in coin flipping	Probability of tossing 7 with a pair of dice
0	1	$1q^4$	0.077	0.0625	0.4823
1	4	$4p^1q^3$	0.276	0.25	0.3858
2	6	$6p^2q^2$	0.373	0.375	0.1157
3	4	$4p^3q^1$	0.224	0.25	0.0154
4	1	$1p^4$	0.050	0.0625	0.0008

hundred rounds of calling heads on the flip of a coin, a player would be most likely to win fifty times, but in a hundred rounds of playing red in roulette he or she would be most likely to win (as we shall see) only forty-seven times. But which forty-seven? Ah, that is the gambler's Holy Grail. And, as we shall see, it most certainly will not be on the first forty-seven. (It may seem peculiar that in a hundred rounds of playing red in roulette you are only likely to win forty-seven times and not fifty, but that comes from the fact that for roulette $p < q$ and so the peak probability is skewed away from the mean.)

We shall never see that gambler's Holy Grail. Yet there are things we can know. Amazing things! From table 8.1 we see some symmetry with both roulette and coin flipping, but much less so with dice. If we picture the column for roulette in table 8.1 as a bar graph plotting the number of times red appears versus the probability of getting that number of reds (see figure 8.1), we see a skewed symmetry about the number of wins being 2, while the center of gravity of the graph (the geometric balancing point) seems to be over a number less than 2. When the number of rounds increases to 8 the skew is even more pronounced (see figure 8.2). If we increase the number of roulette rounds to, say, 100, the graph will look smoother. There would then be 101 rectangles, each having a base one unit wide.[1]

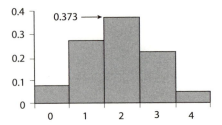

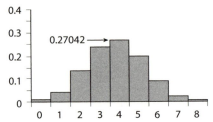

FIGURE 8.1. Probability of winning on red in four rounds of roulette.

FIGURE 8.2. Probability of winning on red in eight rounds of roulette.

Take a look at figure 8.3. It is what is called a *frequency distribution*, because the height at each number of successes tells us how frequently those successes are expected to occur. The bars are distributed over the horizontal axis in such a way that the sum of their areas equals 1. A few points to note. Figure 8.3 illustrates the distribution frequency only between the markings 27 and 67. Though the real distribution graph extends from 0 to 100, most of the graph is concentrated between 32 and 62; beyond those marks, the probabilities are spread out so thinly (almost 0) the beginning and end would hardly be seen at a practical scale. (Below 32 and above 62 the probabilities are so small we cannot see them on the graph. For example, $P(31) = 0.00034$ and $P(63) = 0.0006$.) It seems a good bet that in 100 rounds of roulette,

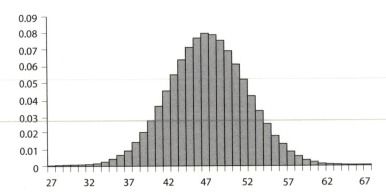

FIGURE 8.3. Probability of winning on red in one hundred rounds of roulette.

red will turn up near 47 times (where the highest bar sits). It also tells us that red is much less likely to turn up 20 times or 80 times.

One more point. As the number of rounds increases, so does the symmetry. In general, for any games where *P(W)* is close to *P(L)*, the curve will become more symmetric as the number of rounds increases, but the peak will shift to the left of center if *P(W) < P(L)*; the greater the difference between *p* and *q* the bigger the shift. So what could be causing such symmetry and what could be causing the shift? For the answer to the symmetry question, we look at the so-called Pascal triangle (see chapter 2).

In the case of flipping coins, where the probability of heads is also the probability of tails, there is perfect symmetry. In the case of shooting craps there is no symmetry and the probability is largest at zero. This is because the probability of getting 7 is far from the probability of not getting 7.[2]

The probability that a roulette ball will land on red *k* times in *n* rounds is

$$C_k^n \left(\frac{9}{19} \right)^k \left(\frac{10}{19} \right)^{n-k}.$$

Why? Look at it this way. We first compute the probability that red is hit on every one of the first *k* spins. Since the probability of hitting red is 9/19, the probability of hitting red *k* times is $(9/19)^k$. However, we also want exactly *k* hits in *n* rounds, so we must also compute the probability of *not* hitting red on the remaining *n − k* rounds. Red must happen again and again *k* times and not happen for *n − k* times; that is, *P(red happening exactly k times in n tries) = P(red)^k P(not red)^{n−k}*. (Remember that the probability of failure is always 1 minus the probability of success. This is because the success and failure are mutually exclusive and the event must end in either failure or success; i.e., P(Success) + P(failure) = 1.) Hence, the probability of not hitting red on the remaining *n − k* rounds is $(10/19)^{n-k}$. Hence the probability of hitting red on every one of the first *k* rounds is

$$\left(\frac{9}{19} \right)^k \left(\frac{10}{19} \right)^{n-k}.$$

In the end, we don't really care whether or not red is hit the first k times or the last k times; nor do we care about the order in which red appears in n rounds. So we must multiply

$$\left(\frac{9}{19}\right)^k \left(\frac{10}{19}\right)^{n-k}$$

by the number of ways k reds can be distributed in n rounds. That number is C_k^n. So, the probability of getting exactly k reds in n rounds is

$$C_k^n \left(\frac{9}{19}\right)^k \left(\frac{10}{19}\right)^{n-k}.$$

Figure 8.3 displays these $n+1$ terms. Note that the height of the middle bar is the probability of getting an even break between red and black, which is computed to be

$$C_{50}^{100} \left(\frac{9}{19}\right)^{50} \left(\frac{10}{19}\right)^{50} = \frac{100!}{50!50!}\left(\frac{9}{19}\right)^{50} \left(\frac{10}{19}\right)^{50} = 0.06928182.$$

Since this number is smaller than the probability of getting 47 reds

$$C_{47}^{100} \left(\frac{9}{19}\right)^{47} \left(\frac{10}{19}\right)^{53} = 0.0795,$$

it is more likely that in the course of a hundred trials, the player will not win often enough to get an even break. How clever of the game designers to throw in a green zero and a double zero.

Let us step back to recall what we really wish to measure. Recall that in figure 8.3 the area of the bar (rectangle) above 50 is measuring the probability that the ball will fall into the red pocket 50 times. To find the likelihood of red having an even or better chance, that is, the likelihood that the ball will fall into the red pocket between 50 and 100 times, we simply sum the areas of the rectangles above and between 50 and 100.

Now the bar above 50 represents the probability that in 100 rounds of roulette a player will break even, that is, win as often as lose. Any bar over a number below 50 will give the probability of finishing 100 rounds at a loss—the lower the number, the bigger the loss. We have

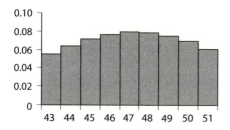

FIGURE 8.4. A modification of figure 8.3.

already calculated that the probability of breaking even is 0.0693. We may calculate the probability of winning k times in 100 rounds by computing

$$C_k^{100} \left(\frac{9}{19}\right)^k \left(\frac{10}{19}\right)^{n-k} .$$

Figure 8.4 is an enlargement of figure 8.3 in the neighborhood of 47. We see that the highest probability on our graph is 0.0795 and that occurs at 47 (see figure 8.4 and table 8.2).

To measure the true probability of ending without a loss, we must sum the areas of all the bars above any number greater than 49, for to do *better* than break even, we would hope to win any number of times as long as it is a number greater than 50. For that, we must take the sum of all probabilities over 50. Such a calculation would be extremely difficult without the aid of a computer or at least some approximation techniques to simplify the calculations, so you might wonder how probability theorists handled such calculations back in the nineteenth century. A simple computer calculation tells us that the sum of the probabilities from 51 to 100 is

$$\sum_{k=51}^{100} C_k^{100} \left(\frac{9}{19}\right)^k \left(\frac{10}{19}\right)^{100-k} = 0.265.$$ (11)

TABLE 8.2
Probabilities Near the Mean

Number	43	44	45	46	47	48	49	50	51
Probability	0.0547	0.0638	0.0715	0.0769	0.0795	0.0790	0.0755	0.0693	0.0611

This is the expectation of doing better than breaking even over the course of 100 rounds. It tells us that at, say, $10 a round and even odds (1 to 1), the average loss over 100 rounds would be $735.

Everything we just said was theory. Until now we have not played real roulette; we have imagined the game played with an ideally spherical ball rolling and bouncing round a flawlessly balanced wheel with precisely spaced pockets in a perfectly steady room in some world we have never seen and that has never existed. That is the theoretical side of what we are about to do. But real wagering takes place in a different world, a physical world where balls and wheels are machined and manufactured to invisibly severe (still imperfect) tolerances by human-made machines built from molecules of atoms with electrons, neutrons, and unnamed subatomic particles. Recall Pearson's collected observations of 4,274 outcomes of a specific Monte Carlo roulette wheel, where the divergence of theory and practice had odds of a thousand million to one.[3] The connection between the ideal and physical is mysterious, and we shall make that connection here.

In the physical world, we could test bona fide roulette wheels for fairness or biases by making a table of observations that could be pictured by a frequency distribution graph. Such a picture may not look anything like the graph of our perfect model, but if the wheel were indeed somewhat fair, and if we were to observe enough rounds, then the graph of observed outcomes would resemble (in shape at least) the graph in figure 8.3. Now, whenever we perform an experiment, we have n outcomes $O_1, O_2, O_3 \ldots O_n$ with respective probabilities p_1, $p_2, p_3, \ldots p_n$. This is the observed probability distribution found by knowing the various ways each outcome can occur. Of course, the sum of the n probabilities must be 1, since we would assume that certainty is distributed among all the possible events. To take tossing dice as an example, any one of six faces can be the outcome, each with a probability of 1/6. If the game is fair (i.e., the roulette wheel, die, or coin is unbiased), then the experimental version of the distribution should turn out to resemble the theoretical distribution, though we should expect some discrepancy, as the wheel or die or coin is not a perfect object in a perfect world.

In this context, *perfect* translates to *mathematical*. How far from perfect are our observations? That is what we wish to determine, and
that shall be done by comparing the data collected by our observations with the data *expected* in a perfect world. You see, there are
gamblers who know that the odds are against them and yet believe
that, not infrequently, the physical world wanders from its expectations to favor a random someone; and inevitably, with the powerful
thought, *someone has to win*, those gamblers are willing to risk heavy
bets against the mathematical expectations of Fortune.

Let's hope that the newer wheels of Monte Carlo are fairer than
they were at the end of the nineteenth century. After all, Pearson's
observations resembled the expected theory and yet the discrepancy
had odds of a thousand million to one! And there is an authentically
verified story that sometime in the 1950s a wheel at Monte Carlo
came up *even* twenty-eight times in straight succession. The odds of
that happening are close to 268,435,456 to 1. Based on the number
of coups per day at Monte Carlo, such an event is likely to happen
only once in five hundred years.[4]

In the first half of the twentieth century (before computers), to
make calculations of expectations for a large number of roulette
coups, mathematicians would rely on the *central limit theorem*, which
gave an ingenious way to approximate the computation by trans-
forming the distribution bar graph to a smooth graph whose areas to
the right of any vertical line were known. Such theoretical approxi-
mations are relics of the nineteenth century, when computations of
C_k^n for k near the range between 30 and 80 would have been quite
difficult. We have some notion of this in *The Doctrine of Chances*, in
which the French mathematician Abraham de Moivre tells us that
though the solutions to problems of chance often require that some
terms of the binomial $(a + b)^n$ be added together, very high values of
n tend to present very difficult calculations. The calculations are so
difficult, few people have managed them. De Moivre claimed to
know of nobody capable of such work besides the two great mathematicians James and Nicolas Bernoulli, who had attempted the problem for high values of n. They did not find the sum exactly, but rather
the wide limits within which the sum was contained.[5] Today that

method is still done as a sentimental habit of the past. There are good reasons for using such methods when dealing with real-life events whose probabilities seem to distribute *normally*.

Figure 8.3 contains 101 bars (rectangles). The tops of the bars mark an impression of an almost smooth curve. Such a curve is called a *frequency distribution curve* because it distributes the frequencies of positive outcomes according to the likelihood p. (The larger the number of bars, the smoother the curve.)

We shall be making a few area-preserving transformations on the curve, and since the areas of the bars represent probabilities, our transformations will continue to preserve probability. Since the base of each bar (rectangle) is one unit in width, the distributions of probabilities are the areas as well as the heights of the rectangles. The transformed graph (figure 8.5) will contain all the information we want to preserve; however, to get a clearer picture of that information we make some simple modifications without altering that information.

First, we shift the entire graph so that the high point is centered at 0. No information is lost, except that we must interpret the meaning of the new graph. It is now the distribution of probabilities of the incremental increase or decrease of reds over blacks. One further modification: We shrink the curve by a factor of 5 in the vertical direction and magnify it by that same factor in the horizontal direction. (The factor of 5 comes from the computation of $\sqrt{Npq}$, where N is the number of rounds—in this case 100—p is the probability of getting red, and q is the probability of not getting red. The precise number is 4.99307. It is rounded to 5 for the convenience of instruction.) Of course the vertical axis will no longer represent probability. That job rests with the areas of the rectangles, and those areas have not changed because we magnified the horizontal and contracted the vertical by the same factor.

What have we achieved? Here's the inspired idea. The bar graph that appears in figure 8.3 may be closely approximated by a simple function,

$$y = \frac{1}{\sqrt{2\pi}} e^{\frac{-x^2}{2}},$$

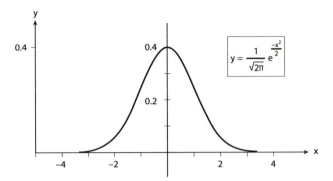

FIGURE 8.5. The graph of $y = 1/\sqrt{2\pi}\, e^{-x^2/2}$, the standard normal curve.

where e is the base of the natural logarithm, approximately 2.1718. The graph of that function appears in figure 8.5 and is called the *standard normal curve*. The important and surprising thing to understand is that this well-studied simple function describes a great many natural phenomena resulting from chance behavior, as long as that function is interpreted correctly.

You may be asking: How can that be? What does this curve—this curve that has no clue that it is modeling events of roulette—have to do with the specific case of balls falling in a red pocket of a roulette wheel? Moreover, you may be amazed to hear that this same curve models coin flipping just as well. The answer is: There is a con, a sleight of hand manipulating the curve that models our particular events by shifting, shrinking, and expanding areas to fit the standard curve, all the while keeping a record of the changes so that no information is lost and that those changes could be translated numerically.

To illustrate the idea, consider the two curves in figures 8.6a and 8.6b. Each has area equal to 1. Each is a transformation of the graph of the function

$$y = \frac{1}{\sqrt{2\pi}}\, e^{\frac{-x^2}{2}}.$$ (13)

Just two numbers, the mean $\mu = 47.9$ (where the curve is centered) and the standard deviation σ (the measure of how quickly the curve

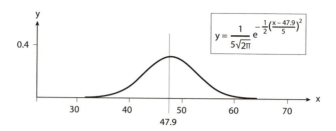

FIGURE 8.6a. Normal curve with $\mu = 47.9$ and $\sigma = 5$.

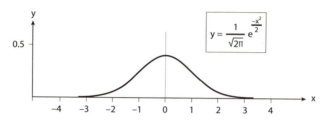

FIGURE 8.6b. Transformation of 8.6a into standard normal curve.

spreads from the mean), determine the transformation from the curve in figure 8.6a to that in 8.6b. In this case $\sigma = 5$.

If the graphs look identical, that's because they *are* identical. Their difference is in their scales; note the numbers on the axes. However, figure 8.6a came from a specific model of betting red in roulette, whereas figure 8.6b is a general function that seems to have no specific connection to betting red in roulette.

In mathematics as in life, rarely do we get something for free. So, in the end, though things seem so simple, we do have work to do to pay for what we are about to gain. We had to first rigidly shift the graph so its center fell above $x = 47.8$, the expected number of times a red would turn up in 100 rounds. We had to compute a scalar (a scaling factor) by which to shrink the curve horizontally and magnify it vertically. The computation of both the shift and the scalar involved knowing something about roulette specifically, namely that the probability of success p (the ball falling into a red pocket) is 9/19. Once we had that specific p we were able to compute the scalar as $\sqrt{Npq}$, where N is the number of rounds, p is the probability of success, and q is the probability of failure $(q = 1 - p)$. In other words, the scalar (the standard deviation) for our particular game of playing red in roulette is

$$\sqrt{100 \left(\frac{9}{19} \right) \left(\frac{10}{19} \right)} = 4.99307 \, ,$$

or approximately 5.

Here is how the theory works. Suppose we have a graph of y vs. x. We can make simple transformations of the variables x and y to new variables X and Y; $X = x - a$, to rigidly slide the original graph a units to the right; $X = x/b$, to horizontally scale the original graph by a factor of b; and $Y = cy$, to vertically scale the original graph by a factor of c. Then we have a new graph, Y vs. X. Suppose we have a frequency distribution curve with p relatively close to q. Then we may transform x into X by letting

$$X = \frac{x - \mu}{\sigma}$$

and y into Y by letting $Y = \sigma y$, where μ is the mean and σ is the standard deviation.

By this transformation every binomial frequency curve is transformed (through shifting and scaling) into the standard normal curve

$$Y = \frac{1}{\sqrt{2\pi}} e^{\frac{-X^2}{2}}$$

whose graph appears in figure 8.6b. A magic trick, where this curve comes out of a hat, is something special and powerful. (The discovery of the curve itself goes all the way back to Abraham de Moivre and Pierre-Simon Laplace. It is what you get from the normal distribution

$$y = \frac{1}{\sigma \sqrt{2\pi}} e^{-\frac{1}{2} \left(\frac{x - \mu}{\sigma} \right)^2}$$

when $\mu = 0$ and $\sigma^2 = 1$.)

The numbers at the base of the curve in figure 8.6b are counting the numbers of standard deviations from the mean. The horizontal axis is marked in units of standard deviation. Heights on the curve are no longer measures of probability, for they have been scaled, shrunken down to preserve the area under the curve. We get something in return for our work. We know the area over the interval $a \le X \le b$ is known for any values of a and b. In particular we know that 68 percent of the area lies above the interval $-1 \le X \le 1$ and 95 percent of the area lies above $-2 \le X \le 2$.

At one standard deviation to the right and left of the curve there are inflection points, where the curve's S-shape moves along from being concave down to concave up or vice versa. One standard deviation for a red outcome on a hundred rounds of roulette is not the same as one standard deviation for heads on a hundred rounds of coin flips. So though the curves in each case are similar in form, their interpretations will be different. Nevertheless, the standard normal curve gives us a single model for both roulette and coin flipping. So, though the curve in figure 8.5 may be the same for many different gambling distributions of odds, keep in mind that the markings ±1, ±2, ±3, and the height at the center must be interpreted by specific calculations of the mean and standard deviation, which will depend on the number of rounds, and the likelihood of a positive outcome.

As N grows, so does the standard deviation $\sqrt{Npq}$. However, the standard deviation is measuring how scattered the outcomes are from the mean. Not only is a larger chunk of the horizontal axis grouped under one standard deviation, but also (for a large number of trials) a good deal of the area under the curve will be considered under a single standard deviation from the mean. In the case of our normal standard distribution, 68 percent of the area under the curve is located between one standard deviation to the left of 0 and one standard deviation to the right of 0.

What does all this mean? We have translated any deviations from the likelihood of success to the area under one standard curve, a curve we know a great deal about. We know the area below the curve and between the markings. We may now ask how likely certain events are: for example, how likely is it for red to appear between 50 and 70 times in 100 rounds of roulette?

Our simple bell-shaped curve gives us a means of computing the chances of winning at roulette a specified amount over a long string of bets. Sure, the odds are slightly in favor of the casino. And here is where roulette differs from coin flipping. In one round of roulette the expectation of winning is 9/19, the probability of the ball falling into a red pocket. Now 9/19 is deceptively close to 1/2, and many novice gamblers think that with an expected value so close to 1/2, they have close to an even chance of winning. It is true that as the number of rounds increases, the distribution graph more and more resembles the normal distribution curve. But something also happens to offset the expectation. If N is the number of rounds played, the peak of the distribution curve will be (by definition of expected number) above the expected number of wins along the horizontal axis. As N grows that expected number will move further and further toward lower values.

Playing a single round does give *almost* 1-to-1 odds, but that *almost*, that slight asymmetry between red and black caused by those two nasty pockets 0 and 00, are keys to casino profits. What happens on the next round and the round after that, or on the fiftieth round? First, we compute the horizontal scaling factor that gets us from our original distribution that was in terms of numbers of rounds to our standard normal distribution that is in terms of standard distributions. That factor is

$$\frac{x - \left(100\left(\frac{9}{19}\right) + \frac{1}{2}\right)}{\sqrt{100\left(\frac{9}{19}\right)\left(\frac{10}{19}\right)}} = \frac{x - 47.87}{4.99}.$$

So $x = 50$ on our distribution curve corresponds to $X = 0.427$ on the standard normal curve. We are interested in the area under the standard normal curve to the right of the vertical line through $X = 0.427$. That area turns out to be 0.3336. Table 8.3 shows the expectations of doing better than breaking even after playing N rounds, assuming each round costs \$10. The amounts shown in table 8.3 may not be impressive to someone who may wish to gamble for amusement. However, that gambler should keep in mind that these expectations are all below an even break of 0.5. In other words, luck—if he or she has any—would surely have to work feverishly hard against all the stochastic forces of the casino—not to mention those of the

TABLE 8.3
Expectations of Doing Better Than Breaking after Playing
N Rounds (assuming each round costs $10)

N	Expectation	Earnings	Cost	Loss
50	.303	$151.50	$500	$348.50
100	.265	$265	$1,000	$735
1,000	.0475	$475	$10,000	$9,525

universe—to walk away with any real profit in the long run. And what do I mean by long run? You might say, I don't stay at the tables long enough to gamble one thousand times. Ah, but remember that that roulette ball doesn't keep a history of past rolls. If you gamble regularly, one thousand rounds happens before you know it, and it doesn't matter whether or not you took a break from gambling for a month or a year. Your personal history accounts for the accumulation of your personal gains and losses—it's always your money added to, or deducted from, net worth.

And here's another thing to keep in mind. Table 8.3 assumes that your bets are fixed at $10 each round. Ha! That's conservative betting. When you have a gain—and remember that gains are already tabulated in the table—the psyche takes over with great force to compel those chips to compound into the next stake. We shall find out how great that force might be in part 3 of this book. If you win $20, you might be tempted to bet the whole $20 in the next round. This complicates the expected earnings. Gamblers who come with a system, such as the Martingale system of doubling bets on every loss, had better have a large pile of cash ready to lose. Imagine starting with a $10 bet and losing ten times—not likely, but possible. Using this system, that would mean a loss of $10,230. You should be ready to shell out another $10,240 on top of having lost your $5,120. If you happen to win on that eleventh round, your total winnings would be $10.

Okay, so now you argue that expectations are averages, and that some gamblers will win big while others lose big, and that the universe of gamblers together have a negative expectation. Others will look at table 8.1 and declare that it is just as likely that heads stay in the lead for a long run of coin flips; after all, the Dow Jones average does eventually go up. Both points are valid; there are winners. If nobody ever won at roulette, no one would be foolish enough to gamble at roulette and the casinos would close. So occasionally someone wins and, less occasionally, someone wins big time. If not, there is always the image of James Bond to keep believers betting against the odds. He looks so calculating, so dashing and effortless in his tuxedo, holding a dry martini, coolly sliding chip piles to his side. There is something about those luxury casinos, like the one in Monte Carlo, that make us all imagine ourselves as 007.

The man on the street has an impression that the house has a large advantage over players. That may be true for some games, such as slots; however, in general the house advantage is, indeed, as we have seen in the case of roulette, slight. That slight positive edge and large bankroll should be enough—for anyone who understands the mechanism of compounding unfavorable odds—to dissuade the amateur gambler from expecting a profit over longtime play. There really is little hope of beating the house in the face of the law of large numbers. Losses from multiple bets against a house with a slight edge can add up fast.

CHAPTER 9

A Truly Astonishing Result

The Weak Law of Large Numbers

When the game of dice breaks up,
the loser, left dejected,
rehearses every throw and sadly learns . . .
—*Dante Alighieri, Purgatorio*

Mathematicians have frequent moments when they are struck by magnificence and beauty. For them, that beauty emerges from the elegance of nature's web of connections: reality with theory, physics and nature with mathematical truth, mathematics with its own networks and meshes of certainties. Jacob Bernoulli had such a moment of discovery when he took Cardano's sketch of an idea to prove the weak law of large numbers. He wrote in his *Ars Conjectandi* (published in 1713),

> Whence at last this remarkable result is seen to follow, that if the observation of all events were continued for all eternity (with the probability finally transformed into perfect certainty) then everything in the world would be observed to happen in fixed ratios and with a constant law of alternation. Thus in even the most accidental and fortuitous we would be bound to acknowledge a certain quasi-necessity and, so to speak, fatality. I do not know whether or

not Plato already wished to assert this result in the dogma of the universal return of things to their former positions [*apocatastasis*], in which he predicted that after the unrolling of innumerable centuries everything would return to its original state.[1]

The remarkable result referred to is what was originally called Bernoulli's theorem and now called the weak law of large numbers (coined so in 1837 by the French mathematician Siméon-Denis Poisson who made significant contributions to probability theory); it is one of the fundamental theorems of probability. Its title may mislead one to believe that it is a physical law; however, it is a bona fide mathematical theorem. The innocent gambler confuses the law with a faith that all events do tend to some recognized a priori probability—for example, the impression that for a large number of coin tosses, the number of heads and the number of tails will verge toward each other. In a very limited sense this is true. But let's be careful. Overly simplified, this is what the theorem says: Suppose we observe a large number of trials in which a particular event is hoped to occur—for example, red turning up in N spins of a roulette wheel. Let k be the number of times red actually turns up. Then we have a success ratio k/N. The question is how close is k/N to $p = 9/19$, the mathematical probability that it *will* turn up? Bernoulli's answer is this: the probability that the success ratio differs from p is as close to zero as one wishes, provided that N can be taken as large as needed to force that condition. (In modern notation, where ε represents any small positive number chosen,

$$P\left[\left|\frac{k}{N} - p\right| < \varepsilon\right]$$

converges to 1, as N grows large.)

The gambler who may have heard of this theorem confuses what it says with the idea that for high values of N the average success ratio converges to p. He or she confuses the outcomes of events with the probabilities of those outcomes. A very big difference! To take coin flipping as an example, the confusion suggests that since $p = 1/2$, the total number of heads will converge to the total number of tails over the long run. No! That is not what the theorem says. Bernoulli's weak law of large numbers is saying that the likeliness of that happening is converging on certainty over the long run. It does not guarantee that that would happen in any individual case. As an example, let us suppose that we have a game where

$$P\left[\left|\frac{k}{N} - p\right| < .0001\right] > .999$$

when N is large. This says that it is almost certain that k/N differs in absolute value from p by less than 1/10,000. But it does not preclude any of the unlikely events from happening often in the game, early on or later. In fact—and here is one other component of the gambler's confusion—even if the success ratio becomes close to p, there is no assurance that it will continue to be close. Moreover, it turns out that a slightly stronger version of the weak law of large numbers tells us that though the success ratio is likely to converge on p, the actual success values tend to behave increasingly wildly. Consider this surprising statement: The probability that the actual number of successes deviates from the expected number kp becomes more and more likely as the number of trials grows very large. Though counterintuitive, it is true.[2]

Figure 9.1 shows the results of an experiment in which a die is tossed 1,000 times. The graph marks the average number of spots on the die over k tosses. The expected value (in this case, the mean) is 3.5. It gives real power to the choice of the name *expected value*; the *mean* is the true expected average in the long run.

So it seems reasonable that the longer the average stays near the mean the harder it is to steer that average away from the mean. For example, by the time $k = 400$ the average is, say, from the looks of the curve, 3.5. Suppose we perform an experiment in which the average is 3.6 and the next 50 rolls are all 6s, an extremely unlikely scenario. Then the average moves up to 3.87 by the time $k = 450$. If the average is 3.6 by the time $k = 800$ and the next 50 rolls are all 6s, then the average is just 3.74 by the time $k = 950$. You get the idea.

The Bernoulli theorem is truly amazing. This theorem is telling us that though nature is so unpredictable because of its unfathomable number of undeterminable variables, we do have fantastically clever

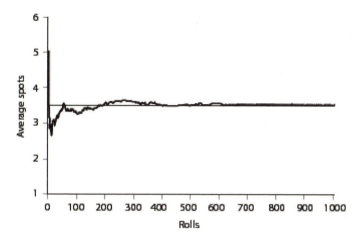

FIGURE 9.1. The results of an experiment in which a die is tossed 1,000 times. *Source*: http://en.wikipedia.org/wiki/Image:LLN_Die_Rolls.gif.

measures of its secrets. Since its publication in 1713, the Bernoulli theorem has gone through a series of upgrades.

The nineteenth-century Chebyshev inequality, named after the Russian mathematician Pafnuty Lvovich Chebyshev, is the mathematical germ of the weak law of large numbers. In modern notation, where X is a random variable, is its mean, and its standard deviation, it states that for any real number $k > 0$,

$$P\left[\left|X - \mu\right| \geq k\sigma\right] \leq \frac{1}{k^2}.$$

It means that we have some real knowledge about the range of X. The probability is concentrated on outcomes that are no more than k standard deviations from the mean. Now this is a fairly weak bound of the range. Nevertheless, when we think of it, it is an amazing statement. We are talking about any string of observations you care to think of! Remember that X represents a large list of possible outcomes $x_1, x_2, x_3 \ldots x_n$, whose probabilities that those outcomes will happen, $P(x_1), P(x_2), P(x_3) \ldots P(x_n)$, distribute over the whole range of outcomes. How can we expect such conformity to a mathematical inequality? It doesn't even matter what the experiment is. If we choose k to be 2, then we know that at least 1/4 of the probability will be concentrated on values further than 2 standard deviations from the mean. So 3/4 or 75 percent of the probability would be concentrated no further than 2 standard deviations from the mean.

If we understand this correctly, we must be stunned, for the theorem is utterly general; it incorporates all possible cases! Not just the ones that have happened in the past and can be modeled by a history of observations, but all future cases that have never happened and may not happen in a billion years, cases that may never be experienced. And here is more magic. The probabilities $P(x_1), P(x_2), P(x_3) \ldots P(x_n)$ may never be known, and yet we can expect that at least

$$1 - \frac{1}{k^2}$$

of the probability will be within k standard deviations from the mean. You may think this is a load of tautological jargon, since knowing nothing about the probabilities leaves us with knowing nothing about the mean and standard deviation. And if we don't know the mean and standard deviation, what good is it to know how much is within k standard deviations from the mean?

Good question. Mathematics has accounted for that. The mean

$$\mu \text{ is defined by } \mu = \frac{x_1 + x_2 + x_3 \ldots + x_n}{n}$$

and the standard deviation by

$$\sigma = \sqrt{\frac{\left(x_1 - \mu\right)^2 + \left(x_2 - \mu\right)^2 + \left(x_3 - \mu\right)^2 + \ldots + \left(x_n - \mu\right)^2}{n}}.$$

Now let's say you are an ancient Persian king rolling an astragal marked with sides A, B, C, and D. You toss it ten times and A comes up 4 times. You toss it again ten times and A comes up 6 times. On the third set of ten tosses A comes up 3 times. And after another set of ten tosses it comes up 7 times. In this particular example (admittedly an imperfect example of very few trials), we have $x_1 = 4$; $x_2 = 6$; $x_3 = 3$; $x_4 = 7$. Clearly, you know nothing about the probability of the success of A. Almost nothing about that probability can be determined from the rough geometry of the astragal. Yet, out of your little experiment of tossing the astragal in four sets, ten times each, we can compute a mean and a standard deviation from the mean.

$$\mu = \frac{4 + 6 + 3 + 7}{4} = 5$$

$$\sigma = \sqrt{\frac{(4-5)^2 + (6-5)^2 + (3-5)^2 + (7-5)^2}{4}} = 1.58$$

Without knowing anything about the particular probability that your astragal will land on side A, you know that 3/4 of the time your values will be within 2 standard deviations from the mean. Since we know the mean to be 5 and the standard deviation 1.58, we know that 3/4 of the time (on average) the values will fall between 1.84 and 8.16. Indeed, this is true, for 100 percent of the values of our little experiment are within those boundaries.

It seems that Chebyshev's beautiful little inequality is a bit loose around the edges and perhaps deficient for making useful calculations. It is. But computing percentages of success was never its goal. Its intent was to tackle the problem of quantifying uncertainty, a problem that started two hundred years earlier with the legendary correspondence between Fermat and Pascal and again at the beginning of the eighteenth century when the subject of the weak law of large numbers was splendidly advanced by the publication of Bernoulli's *Ars Conjectandi*. And yet there must be something more to it, for with it we can prove the weak law of large numbers and *that*, we all must admit, gives us a commanding handle on uncertainty (see appendix C).

Now this is a truly astonishing result. Very roughly speaking, it says that in the long run the difference between the actual mean (which of course is entirely unknown before the events occur) and the mathematically calculated mean is likely to be as small as one wishes, provided that the number of trials *N* is large enough. How is it possible for random events (with absolutely no history of each outcome) to have a mean close to a mathematically calculated number?

The amazing thing here is that all this is simple algebra and yet we have confined the uncertainty of any real-life binomial experiment.

How did the phenomenology of the experiment enter the mathematics? Amazed? You should be! It is a priceless piece of intellect. The steps in this proof, as well as those in Bernoulli's original proof, are entirely mathematical. I should point out that Bernoulli's proof, which is a bit more difficult to follow, rested on using both combinations of n things taken k at a time and expansions of powers of binomials. It identifies certain terms of those expansions with possible outcomes of experimental events. If one looks carefully at the five lemmas Bernoulli presents to prove his theorem, one is convinced that they are purely mathematical and should have had nothing to do with the random winds of Fortune.[3]

Let us deconstruct how the connection between mathematics and the random winds of Fortune came about. We start with the question of how some simple algebra could possibly tell us so much about the secrets of chance. What's the connection? Surely, algebra can tell us about determinant phenomena of the real world—the structures of bridges and dams that obey mathematical computations. Planes fly and windows break according to mathematics. Glass breaks at certain resonant frequencies; plane airfoils lift when the pressure above is lower than the pressure below. But when it comes to chance, the connections seem far more mystifying. Dice? How could we possibly know which way they will fall on any given throw?

Look carefully at what we have done in appendix C and you will see that all the mathematics connecting Chebyshev's inequality to the weak law of large numbers rested on one key ingredient: the simple definition of *expected value*. The credit for the idea of expected value goes partly to Fermat and Pascal and partly to the seventeenth-century Dutch mathematician Christiaan Huygens, who, after recognizing the contributions of his eminent colleagues, began his *De Ratiociniis in Ludo Aleae* (On reasoning in games of chance) with the following:

> Although the outcomes of games that are governed purely by lot are uncertain, the extent to which a person is closer to winning than to losing always has a determination. Thus, if a person undertakes to get a six on the first toss of a die, it is indeed uncertain

whether he will succeed, but how much more likely he is to fail than succeed is definite and can be calculated.[4]

In this quote we have the first printed recognition that the phenomenon of uncertainty recognizes the difference between number of successes and the *likeliness* of the number of successes. The *De Ratiociniis* (1657) itself is the earliest published work on probability.[5] Huygens goes on to say,

> Similarly, if I were to contend with someone on the understanding that three games are needed to win, and I had already won one game, it would still be uncertain which of us would win three games first. Yet we can calculate with the greatest certainty how great my expectation and my opponent's expectation should be appraised to be. From this we can also determine how much greater a share of the stakes I should get than my opponent if we agree to quit with the game unfinished, or how much should be paid by someone who wanted to continue the game in my place and with my lot.[6]

He gives this example: It is a game of chance, but you have to pay to play. A person hides three coins in one hand, seven in the other and offers you the coins of whatever hand is chosen. How much should you pay to play? The answer is 5, that is, the expected value or average of 3 and 7. His first proposition, in essence, says that if either a or b could equally occur, then the expectation should be worth $(a + b)/2$. It is not at all clear that Huygens understood the remarkable power his notion would have on the world.

Expectation is the harness that reins mysterious uncertainty. Our neat little standard deviation, that measure of the amount of scattering from expectation, binds mathematics to the stochastic world. These are the nuts and bolts of probability. Miraculously, from them and simple algebra, we have if not direct management at least a soft measure of phenomenological chance by way of the weak law of large numbers. How extraordinary! That a bridge will survive a thousand years of traffic and wind is a matter of mathematics, geometry, physics, and the molecular testing of the strength of materials, yet the cumulative average of the values of a thousand tosses of a die (3.5) is not at all a matter of the strength of the die. In the ideal world, the physics (velocity, trajectory, air currents, gyro-effects, angular momentum, impact, etc.) surely does have a great deal to do with the outcome. We are reminded of the recent coin-tossing experiments showing that a real penny is entirely determined by the physics of the spin and influenced by imperceptible human controls and tendencies that make its behavior seem random.[7] However, the weak law of large numbers does not come to us from the real world where air currents, gyro-effects, and impact matter, where every throw of a die is influenced by a huge number of forces and circumstances that are hardly measurable, but rather from the ideal.

Concerned with a priori knowledge of chance, where probabilities are known through ideal mathematics, Huygens understood that the nucleus of the theory of probability is simply the expected value. It would have been wholly premature for a mid-seventeenth-century mathematician to know the real truth: that all of nature's random performance, including the behaviors of annuities, insurance, meteorology, and medicine, as well as games of chance, are compelled by expectation. How fortuitous that someone at that stage in the history of probability had thought of the mathematical computation of expected value without knowing that it would turn out to be the most natural measurement of central tendency, the tendency for events to favor the mean.

Like all early works on probability, the *De Ratiociniis* concerned itself with a priori computations. Given a die, it could fall on one of six faces, so the probability p of falling on any particular face is $1/6$. Subsequently, Jacob Bernoulli's *Ars Conjectandi* was a departure from the eighteenth-century view that anyone could count the cases and take the ratio of the number of occurring cases to the total number of all possible cases.[8] Bernoulli asked the question differently to include problems involving disease and weather. He gives an example that is similar to the one Lenin, my probability teacher, used in his class. Bernoulli writes of an urn filled with three thousand white tokens and two thousand black and tells how to find the ratio of white to black if we do not know that there are three thousand white tokens and two thousand black. Blindly choose one token, record its color, and put it back. Repeat this blind picking a sufficient number of times to ensure a pretty good likeliness—a probability that can be made as close to certainty as we wish by increasing the number of picks—that the number of white to black will be in the same secret ratio as three to two.

With a probability as close to certainty as we wish, we can now expect something from a series of independent trials with probability of success p: the relative frequency of success will differ from p by less than any given number $\varepsilon > 0$, no matter how small, provided the number of trials taken is sufficiently large. That is what the weak law of large numbers guarantees.

In theory, the weak law of large numbers should have been intellectually explosive, a tour-de-force mathematical measure of uncertainty; yet, from a practical standpoint, Bernoulli himself may have been discouraged. *Moral certainty* was Bernoulli's term, his requirement that the observational ratio be within one in fifty of the true ratio with guaranteed odds of 1,000 to 1, an extremely harsh criterion even by today's accepted standard of certainty.[9] Because Bernoulli died before finishing the *Ars Conjectandi*, we have an abrupt end after the first example. The numbers are disappointing. After 25,550 observations, where the expectation (unknown to the observer) is 3/5, the chances are 1,000/1,001 that the relative frequency of successful cases would fall within 1/50 of the true expected value of 3/5. With such a large number, after all, what practical advantage can the theorem offer? (This large number is a result Bernoulli's conservative requirement that certainty should be higher than 1,000/1,001, which puts it well over 99 percent. This is far higher than the 95 percent requirement accepted in science today.)

Not much. No dependable information can be determined from a sensible number of experiments. In his book on the history of probability, Stephen Stigler notes that 25,550 would have been considered astronomical at a time when the most up-to-date star catalogue listed just 3,000 stars and the entire population of Basel was then smaller than 25,550.[10] Still, the theological, ontological, and metaphysical implications are massive. On the functional side, there would be other steps along the quantification of uncertainty journey.

In his book *The Theory of Almost Everything,* Robert Oerter writes that the goal of physics from Newton's time to the early twentieth century had been to predict the future.[11] Bernoulli's weak law of large numbers promised to promote that goal. Yet it was the eighteenth-century mathematician Abraham de Moivre who was to make the next move. There is a deep connection between the weak law of large numbers and the central limit theorems we talked about in the previous chapter.

De Moivre noticed that as the number of trials increased, the shape of every binomial distribution (the distribution of n independent yes/no type experiments such as coin tosses) approached the contour of one particular smooth bell-shaped curve. By 1733 De Moivre had found a normal approximation to the binomial distribution in his endeavors to sharpen Bernoulli's attempt at quantifying uncertainty. The problem all along was how to perform a vast number of necessary computations. They would not have been so difficult at a time when mathematicians were used to spending long days trudging though mindless calculations, but terms got terribly large near the middle of the distribution, especially when the number of trials was large. For example, the middle term middle bar of figure 8.3 (the number of rounds it takes to get an even break between red and black in roulette) is

$$C_{50}^{100} \left(\frac{9}{19} \right)^{50} \left(\frac{10}{19} \right)^{50},$$

a very nasty thing to calculate without a trick or some sort of computer. With insightful analysis of the binomial distribution, De Moivre was able to give a more precise, as well as satisfactory, estimate of the terms of the binomial distribution.[12] Surprisingly, it was then possible to compute the binomial probability that in an experiment involving n observations the observed mean would fall within a certain range (a distance measured by multiples of $\sqrt{n}$) of the expected value. He takes an example where $n = 3,600$ and asks how likely it would be for an event to happen more often than 1,750 times and less often than 1,850 times. In that case the range would be between $1800 - s\sqrt{n}$ and $1800 + s\sqrt{n}$. Therefore, since $\sqrt{n} = 60$, we have $s = 5/6$. We can find that probability by finding the area under the normal distribution curve between $-5/6$ and $+5/6$. What is $\textcircled{17}$ remarkable here is that so many natural phenomena obey frequency distributions that may be approximated by a normal curve.

The surprise is that De Moivre's contemporaries missed the significance of this magnificent breakthrough; the work was not applied for some time even after its third publication in 1756 and waited a quarter century for Laplace's *central limit theorem*, which in essence said that in a sample population of size N, with mean μ and standard deviation σ, the sampling distribution of the mean approaches the normal distribution with mean and standard distribution σ. Moreover—and this is the significant point—regardless of the shape of the parent population, the sampling distribution always approaches the normal distribution as the sample size N increases. But that's another story.

Bernoulli—who, like most early determinists, saw God as the ultimate custodian of chance—has given us a method of finding the expected value from no a priori information. The power behind this is enormous, for it applies to the uncertain behavior of nature as well as to games of chance. For example, Bernoulli understood that if we think of the human body as the urn and the tokens as germs and diseases, we could determine by observation the likelihood that disease will enter the body.[13]

Fortune will never admit what she truly knows as fact: that the outcome of a rolling die is not intrinsically random but only seems so in ignorance of the details such as hidden variables (launch angle, friction, etc.) that determine the outcome.[14] Like dice, most phenomena in our universe (especially those affected by atomic authorities) have far too many hidden variables for mathematics to predict their outcomes. We are generally ignorant of the minutiae of such wonders. Yet we have this marvelous gift that was a secret up until the end of the seventeenth century when Bernoulli discovered the law of large numbers to give us a clue that the key to understanding randomness—as well as the means of predicting the future—is in the expected value, that number that almost all random behavior favors in the long run. The law of large numbers celebrates randomness while, in the long run, marking rigid determinism. How magnificent! Most happenings of the universe obey the law of large numbers, even though each individual event has no history of its past. Whether God plays dice or not, long-term trends of expectations are predictable and—almost always—assured.

♣ CHAPTER 10 ♣

The Skill/Luck Spectrum

Even Great Talent Needs Some
Good Fortune

The most fundamental principle of all in gambling is
simply equal conditions, e.g. of opponents, of bystanders,
of money, of situation, of dice box, and of the die itself.
—*Girolamo Cardano, Liber de Ludo Aleae*

There are essentially eight standard gambling games: roulette, craps, slots, lotteries, blackjack, poker, horse racing, and sports.[1] They are the Big Eight. By far, the biggest is lotteries. According to recent surveys conducted by Gallup, 49 percent of Americans spent on average $184 on lottery tickets in 2004, and the hunch is that that figure had increased substantially during and after the 2008–9 recession. Forty-one states now have lotteries earning more than $52 billion, gross. Of course there are plenty of other games to play— baccarat, bingo, Caribbean stud, three-card poker, faro, piquet, keno, video poker, and so on. The list is endless. Most other games turn out to be, more or less, variations on games of the Big Eight.

No one can rationally claim—though some people do—that the outcome of rolling dice is determined by skillful throws or that the landing of a roulette ball is in the skill of the croupier's wrist—unless, of course, cheating is going on—though I should mention that back in 1896 Poincaré analyzed the spin of a roulette ball and found that the outcomes become more and more uniform when a ball is spun with more and more vigor.[2] There are many ways to cheat and new ones are being invented every day—magnetic or loaded dice, or irregularly marked dice in the hands of pros, could be switched in mid-game. Games such as poker, horse racing, backgammon, and blackjack involve varying degrees of both luck and skill, including cheating with marked or "stripper" decks, where non-favored cards are imperceptibly shorter than favored.

Games of skill also require luck. A favored horse can slip in the mud and lose; witness Big Brown, who was heavily favored to win the 2008 Triple Crown, or Mine That Bird, the 50-to-1 long shot who won the 2009 Kentucky Derby. In the searing heat of Belmont, he came in last. Who could have foretold that Da' Tara, a 38-to-1 long shot, would win the race? Talk to Nick Leeson, the rogue derivatives trader who, in 1995, brought down Barings Bank—the oldest investment bank in England—by gambling in the futures market and losing £850 million ($1.3 billion). His unsupervised and unauthorized speculations might have gone very well had it not been for the Kobe earthquake. It was Leeson's high-stakes poker game.[3] Leeson was playing poker with short-term futures on the Singapore and Tokyo stock exchanges, betting that the Japanese stock market would be secure. However, early the next morning (January 17) the Kobe earthquake hit, sending Asian markets into a tailspin. Chasing his losses, he made a series of increasingly risky investments, betting that the Nikkei Stock Average would recover. It didn't. Like many gamblers who chase their losses, he continued to sink into deeper trouble.

If you're going to gamble, you'd better know the risks. In poker, those risks are calculable. See callout 18 for suggested ways to make those calculations. Try calculating the odds of a flush, a straight, and two pair.

(18)

In an actual five-card draw poker game, the master poker player knows the math facts of table 10.1 cold, with no need to make those calculations. However, in the dynamics of an ongoing poker game there are intuitive calculations that must be made before making decisions to drop (show your hand), call (pay the ante), or raise

TABLE 10.1
Odds of Various Hands in Five-Card Draw Poker

Hand	Probability	Odds
Royal flush	0.00000154	1:649,739
Straight flush	0.0000139	1:72,192
Four-of-a-kind	0.00024	1:4,164
Full house	0.00144	1:693
Flush	0.00197	1:508
Straight	0,00392	1:254
Three-of-a-kind	0.0211	1:46
Two pair	0.0475	1:20
One pair	0.423	1:2
No pair	0.5	1:1

(raise the ante). Given a hand, what are the odds of drawing a better hand than an opponent's by drawing the next, say, two cards? The master may not be able to answer with an exact number, for that would require some thorny calculations to be made on the fly, so he or she must rely on the acute sensitivity of an estimate. This must be done on top of the enormous number of behavioral reckonings that enter, as well as the usual performance of acting and bluffing. The mathematics may be thorny when attempted in the middle of a poker game, but they are clear when one has the luxury of pencil and paper. The master is likely to study these specific situations like the sprinter trains for a race.

Let us suppose you have three cards of the same suit, say diamonds, and are hoping for a flush by drawing two cards. The probability of your next card being diamond is 10/47 (that's because there are 47 cards left in both the deck and your opponent's hand, and you have already been dealt 3 of the 13 diamonds). The probability of getting a diamond on the second draw is 9/46 (given that you drew a diamond on the first draw) and so the probability of a flush is

$$\left(\frac{9}{46}\right)\left(\frac{10}{47}\right) = 0.042 .$$

It may seem as if we are not taking into account the cards that were dealt to your opponent. Those cards, whatever they are, will not be dealt to you; they are, in essence, not in the deck—there are actually only 42 cards left in the deck. However, we are assuming you know nothing about your opponent's hand and, therefore, also assuming that your next card will be any of the 47 cards unknown to you. In the middle of a game, some cards may be exposed. That would alter the calculations considerably, but that's what makes poker such a dynamic game.

If you think your opponent has two queens and is about to draw two cards in hope of three-of-a-kind, you could calculate his or her chances. The probability of getting a queen on the first draw is 2/47. There are two possibilities: If your opponent does not get a queen on the first draw then he or she will still have a chance of getting one on the second. So the probability of your opponent getting three-of-a-kind is P(queen on first draw) + P(no queen on first draw)P(queen on second draw). This turns out to be

$$\left(\frac{2}{47}\right) + \left(\frac{45}{47}\right)\left(\frac{2}{46}\right) = 0.084.$$

Of course, things are more complicated by the fact that you may already have a queen or two queens, which changes your opponent's probability of drawing another queen.

Lotteries are pure luck gambles. Whenever I hear of someone wasting money on a lottery ticket, I'm reminded of teenage times spent at an amusement park in Palisades, New Jersey, where there was a wheel marked with 25 numbers waiting for players to pick a number for 25¢ a turn to win $6. A sign above the wheel claimed that the payoff odds were better than even. Better than even! I had a friend by the name of (can you believe it?) Alowishious Dill—Allie for short—who was clever with math ideas. He could clearly explain confusing mathematical notions. Fearing that I was a sucker for better-than-even odds, he promptly tried to enlighten me, claiming that I might just as well have given the attendant my quarters than play the game.

"Suppose that the game offered 23-to-1 payoff odds against picking one number in 25 tries," he said. "There are 25 chances to lose and 1 chance to win. It would cost $6.25 to play 25 times.

"I should be able to win at least a few times in 25," I reasoned.

"Well," said Allie, "that's as close to better than break-even odds [24 to 1] as anyone can get. Right?"

I agreed. Allie figured that I could have just as well put a quarter on every number because every number had the same chance as any other—assuming that the wheel was fair. By betting on every number, I would have been sure to win. However, I would have collected just $6.00, including the winning quarter.

"Well," said Allie once again, "why not just give the man a quarter?"

I didn't understand until years later. For one thing, I hadn't fully understood the equivalence between twenty-five different plays at one quarter a play and the single play with a stake of twenty-five quarters; it was hard to grasp the concept of the wheel's fairness and of how it applied. For another, I had counted on that thing everyone called *luck* to slyly counter the true mathematical odds bias; after all, I would have had twenty-five chances, so shouldn't I have won at least a few times? Why not risk it?

Allie was giving me a lesson in the law of large numbers long before either one of us had ever heard of such a law. Surely luck would come into play in the first twenty-five plays, but what about in fifty, a hundred, ten thousand? In ten thousand spins of the wheel, the true odds would predict a loss of $100. The hard thing to grasp is that luck could have given a net profit in the first five hundred spins, but, in the long run, luck would have been overcome by forceful odds as well as by the firm determination of the law of large numbers.

Payoff odds of 23 to 1 are pretty close to fair. That's what makes it so powerful a weapon against the combined psychological forces of luck and risk. Had the sign above the wheel said 5 to 1, nobody but a fool would have played the game. It would have meant losing on average 19 cents for each quarter staked! If the payoff odds were 25 to 1, things would have been different. One penny would have been gained for every quarter staked.

Still, those commanding psychological forces are somewhat rational; they whisper the basic strategy of all gambling: *maximize expectation while minimizing risk.*

When the odds are just shy of fair, as they are in some table games, luck (whatever it is) can compete, at least in the short run. However, when the odds are ridiculously in favor of one side over another, as they are in some games such as lotteries, luck will almost always be defeated, even in the short run. Lottery odds are so skewed toward the improbable that they give the least fair odds of any gamble.

Consider the Tri-State Megabucks, involving Vermont, New Hampshire, and Maine. Draws are held twice a week. Players purchase a single ticket for $1.00 and mark six numbers from 1 to 42.[4] Six numbers are randomly drawn, without replacement. The odds of winning any prize (picking the correct 3–6 of the drawn numbers) are 21 to 1; the odds of picking the smallest prize (picking three numbers and winning back your dollar) are 40 to 1. (This means getting 3 hits out of 6 possible hits and picking the remaining 3 numbers from 36, or

$$C_3^6 \cdot C_3^{36} = \left(\frac{6!}{3! \cdot 3!}\right)\left(\frac{36!}{33! \cdot 3!}\right) = 142,800.$$

The total number of combinations of picking all six numbers is 5,245,786. Therefore, the probability of picking the correct three numbers is

$$\frac{142,800}{5,245,786} \approx 0.027,$$

giving odds of roughly 36 to 1.) The odds of winning the jackpot (picking all 6 numbers out of 42) are 5,245,785 to 1 (a probability of 0.00000019), less likely than being struck by lightning sometime in a lifetime on a Tuesday morning. (There is one and only one way to hit the jackpot and that is to pick 6 numbers from 42 choices. There are

$$C_6^{42} = \frac{42!}{35! \cdot 6!} = 5,245,786$$

different ways to pick 6 numbers from 42 numbers.)

TABLE 10.2
Megabucks Payoff Odds

Match	Amount	Odds	Probability	Expected value
6 Numbers* (no Bonus)	Jackpot**	1:5,245,785	0.00000019	0.19
5 Numbers + Bonus	$10,000	1:877,192	0.00000114	0.011
5 Numbers*	$1,000	1:24,980	0.00004003	0.040
4 Numbers + Bonus	$50	1:9,991	0.00010008	0.005
4 Numbers*	$40	1:585	0.00170648	0.068
3 Numbers + Bonus	$5	1:440	0.00226757	0.011
2 Numbers + Bonus	$2	1:52	0.01886792	0.037
3 Numbers*	$1	1:39	0.025	0.025

* Not including the Bonus Number. (The Bonus Number is a seventh number drawn after the first 6 distinct numbers are drawn. If there is a prize and the Bonus Number matches one of the 6 numbers played, the prize is increased by a specified amount.)

** Depends on the number of tickets sold and how many weeks have gone by without jackpot winners.

Why would anyone play, when it is the gambling game with the most negative and severe expected value? With a jackpot prize of, say, $1,000,000, the expectation of winning the jackpot is a mere 19 cents. However, it is possible to win one of the other seven non-jackpot prizes. So we must add an expected value of 20 cents (the total excluding the jackpot from table 10.2) to the jackpot expected value, thereby making the expected value for winning any prize 39 cents. For each dollar played the player is throwing away 61 cents.

Hold on! It's worse than that. Keeping in mind taxes and the possibility of sharing the prize (with another winner), the expected value shrinks to approximately 26 cents. In effect you are throwing away 74 cents on every ticket you buy. Other unpredictable factors make things even worse. For example, the pool of players increases with the size of the jackpot, and so the likelihood that a winner will share the jackpot increases.

Take the case of a state with a small population. Vermont generates revenue of over $100 million a year on lottery tickets, with a profit of about $23 million (yet the entire population of Vermont is only 624,000 and the number of households is only 241,000). A 2008 survey conducted by the Center for Research and Public Policy concluded that 40.5 percent of Vermonters—half of whom attended college—played the lottery regularly. On average, households of Vermont spent $414.93 a year on lotteries. That may not be much to some people, but there are those—those who spend more than they can safely afford—maximizing expectations as well as risk.

In Vermont, all lottery profits go exclusively to support the Vermont Education Fund. You'd be better off just donating your dollars to that fund directly—that way you would have a tax advantage over the price of a ticket; it would be an expected value equal to that of buying lottery tickets. On the spectrum of gambling game expectations, from abysmal to fair, lotteries win the abysmal slot. Blackjack and poker, the most frequently played games in America, are fair games for those who know how to play.

Till now we have not examined betting on races. Horse-race betting in America is pari-mutuel; that is, the total amount wagered is pooled and—after a deduction for *vigorish* (the track's take) to guarantee the house edge on the action—divided among those who have staked their bets on the winner. Each gambler is betting against other gamblers, not the house. Electronic ticket machines instantly compute the winning odds and payoffs, which are displayed on a tote (*totalizer*) board. Windows are open for fifteen minutes before each race so bettors may wager $2.00 up to make straight bets of win, place (finish first or second), show (finish in the top three), or a smorgasbord of combinations and choices.

Handicapping relies mostly on subjective odds. With legal pari-mutuel betting, the starting odds are by and large determined by the amount of money bet on a horse and hence on bettor confidence. At post time, the betting public relies on pre-race odds, that is, on ticket sales indicating how bettors are feeling about the horses. There is no mathematical blueprint. The more involved story is this: Post position, past performance, and jockey and track conditions are considered, but no serious mathematical recipe is established to account for how any particular horse feels on the hour of a race. Odds displayed on the tote board start from the racing programs and those who work with the horses or the track (morning-line odds makers). Those people give the tentative odds by handicapping each horse in each race forty-eight hours before post time. Public handicappers who work for the newspapers make their choices after the morning-line odds makers have posted. Finally, the general public gets into the act by placing its bets.

In almost any dozen races, it will turn out that, on average, the favorite does *not* finish first. Of course, that's the way it should be—after all, it is a sport and we recall that someone once said *every time a rack of pool balls is broken, there is a new game.*[5] However, though on average those favorites do not win, they do have better average finishing positions than non-favorites, so there is something to be said for handicapping (personal judgment) after all.

In pari-mutuel betting, the odds at post time do not directly determine the payoff. If the tote displays 3-to-1 odds for the winner, don't expect anything like $6 for every $2 bet on the winner. Indirectly, of course, the tote odds play a major role in that they influence how much money is pooled in each category—win, place, or show. Bets pooled on the favorite will undoubtedly be larger than those on a long shot. And the pool of bets to place will be smaller than that of bets on the winner.

The payoff on a $2 bet is relatively simple to calculate, if we ignore a small calculation of rounding down to the nearest nickel. Let T be the total number of dollars bet to win, W the total bet on the winner, and r a percentage taken out by the track and state.[6] Then, as we shall see in a moment, the payoff for a $2 bet would simply be

$$2(1-r)\frac{T}{W}.$$

However, rounding down to a nickel when dealing with large sums can have significant advantages for the track. We shall see this in the rationale for this strange payoff.

First, the track and state deducts from the total for its take. That leaves $(1-r)T$ in the win pool. Next, W is deducted, leaving $(1-r)T-W$ to be divided among the winners. So each $2 bet should pay

$$2 \cdot \frac{(1-r)T - W}{W}$$

in addition to the original $2, or

$$2(1-r)\frac{T}{W} \text{ dollars.}$$

The problem is that the track deals with real dollars and real cents. The quantity

$$\frac{(1-r)T - W}{W}$$

is likely to be a decimal with non-zeroes beyond the hundredths. For example, let's say the total amount bet to win is $300,000, that $200,000 is bet on the winner, and that $r = 0.15$. Then the track takes $45,000 and computes that the dollar payoff on the winner is 27.5 cents. The track would round this payoff down to the nearest nickel (*the breakage*) and so the dollar odds would become 25 cents. Each $2 bet would have a payoff of $2.50. The track would keep the breakage, $.025 × 200,000 = $5,000. So, adding up: the track and state gets $45,000, the track gets an additional $5,000, and the group of winners gets $250,000 for a total of $300,000, the total amount bet.

It may seem as if we've forgotten about bets to place, to show, and so forth. Not really; those bets are pooled separately. The computations are slightly different. The dollar paybacks for place bets are similarly computed except for the fact that the bettors of two horses will have to share the winnings in proportion to the sizes of the bets on each horse. Let's say that T is now the total amount bet to place and that P_1 and P_2 are the amounts bet for horse 1 and horse 2 to place, respectfully. Then the per dollar payback for horses 1 and 2 are, respectively:

$$\frac{(1-r)T - (P_1 + P_2)}{P_1} \text{ and } \frac{(1-r)T - (P_1 + P_2)}{P_2}.$$

A similar formula gives the per dollar payback for bets to show.

In a strange way, this links up with the law of large numbers, for it seems that, at least for horse racing, large betting groups are fairly good judges of performance. In fact, as the betting group grows larger, the post-race results get closer to the pre-race odds. It's as if the public brings a certain group intelligence into play. The individual bettors are making individual judgments that are suspiciously subjective, yet, when compounded and averaged, their collective results tend toward a baffling accuracy that outperforms the forecasts of individual experts.[7] This notion of crowd intelligence sprouted from a classroom experiment performed in 1920 by the Columbia University sociologist Hazel Knight. Knight asked each student to estimate the classroom temperature, then averaged the estimates to obtain 72.4 degrees. The actual answer was 72 degrees. Granted, it was not an impressive experiment, but its idea did, in later years, lead to more serious research on group intelligence and random sampling.

In the early 1980s when Jack Treynor was professor of finance at the University of Southern California, he conducted a somewhat more impressive experiment. Treynor (now president of Treynor Capital Management) meant to defend the economic theory of free-market capitalism, but his experiment inadvertently supported a more general theory of group intelligence. His illustrious jellybean experiment, which has been replicated many times, had fifty-six students estimate the number of jellybeans in a jar containing exactly 850 jellybeans. The average estimate was 871.[8]

In 1970, a study compiled statistics on the starting odds vis-à-vis average finishing positions for 1,825 races of a season at Aqueduct and Belmont.[9] It turned out that over the entire racing season whenever the betting crowd gave the favorite 2-to-1 odds, the favorite won, on average, one time out of three, excluding the occasional long-shot upset. In other studies as well, the favorite tends to win in the long run of a season about 1 time in 3.

Bets on favorites with very good odds generally have positive expected values. Favorites, however, don't pay well enough, and some studies have shown that betting favorite at post time generally leads to a loss over time, so bettors—especially those wagering with bookies—tend to go for more risky picks. What's perplexing is the *favorite long-shot bias*—that bettors bet on long shots to win even though

long shots rarely win and bettors don't bet on favorites even though
favorites often do win. The phenomenon was first studied in 1949
and found to hold at racetracks worldwide with few exceptions.[10]
Favorites are favorites for good reason; they are much better bets
than long shots. Curiously, consistent betting on long shots yields on
average a loss almost three times greater than losses from bets on
random horses.

Back in the days before 1968, when off-track gambling was illegal in most American cities, bookies would use the favorite long-
shot bias, the crowd's intelligence, and the law of large numbers in
the safety of starting odds by offering high paybacks on long shots
and even money on favorites. In 1960 the number of illegal bookies
in America numbered between 40,000 and 60,000, taking in a net
profit of $2 billion (about $14 billion in today's money). Thirty million Americans made at least one $2 bet at a horse race in that year.[11]
Bookies would keep all the money not staked on the winner, though
they did have expenses, including a hefty amount for *ice* (protection
money). It meant taking some small risk in tempting the bettor with
the high payback of long odds on a long shot against the favorite.
Bookies offered odds with care. If the long shot came in, any gains on
favorites had to be offset by losses on long shots. Fortunately, there
was almost always a gain when a long shot won because—in spite of
the favorite long-shot bias—more money was placed on the favorites
and there were always more long shots than favorites. In fact, the
larger the betting pool, the larger the favorite long-shot bias. So, with
a sufficiently large betting pool, the bookmaker could rely on his
profits with near certainty.

By the time I was ten years old in 1952, lotteries had all but disappeared in favor of its child, numbers, the game where players pick
three or four digits to match those that will be printed in a newspaper the following day. The winning number would generally be a
number nobody could know in advance, a number such as the last
four digits of the daily balance of the United States Treasury. My
uncle Sam brought me to the bingo hall down the street from where
my grandparents lived, the same place where my enterprising uncle
Herman used to sell cold drinks in conical cups. Up a creaky set

of wooden steps, in a large smoky room lit only by bulbs hanging from wires in the ceiling, there stood a man marking numbers on a chalkboard. Sam introduced me as his smart little nephew to a pockmarked-faced, hard-looking character named Red who—as I can still lucidly picture—pulled out a roll of bills and, with a quick lick of his thumb, counted out a few tens to hand over to my uncle. I had seen Red before. He came up to my grandparents' apartment to collect a debt from my uncle Sam. Years later I learned that many of the innocent shops on Ocean Avenue were actually policy shops, taking bets on horses and numbers, and even had craps games running in a back room after hours. The liquor store downstairs was one. The boss was a Damon Runyon character—short and fat, at all times clutching a rumpled copy of the Daily Racing Form folded under his arm. Sam brought me there too to get his cut on a tip about a horse at Belmont. Sam got word from my uncle Al, who got it from his fiancé's father, who knew the owner—I never knew the full story but assumed that something was fixed.

Blackjack, on the other hand, has a well-described mathematical recipe that gives a 3 percent advantage over the casino to anyone who has studied the game carefully. The aim of the game is get closer to 21 than the dealer's hand without going over 21. All picture cards count as ten and an ace can count as either 1 or 11 at the player's choice. According to Edward Thorp—the mathematician who, in 1962, first published his strategy for winning at blackjack—the house advantage in blackjack is between 5 and 8 percent.[12] Now we know that the side with the advantage, no matter how small, will win *in the long run*. With just a 5 to 8 percent advantage, the casino is able to pay its enormous overhead bills as well as lavishly support its owners or shareholders. Think about what percent a single player might need to have a happy payoff. What if the advantage could be tilted toward the player with an advantage of just 3 percent? After all, the player has no overhead, other than food and a hotel room, which is likely subsidized by the casino anyway.

Cheating aside, advantage can be shifted in several ways. First, we are aware that the composition of the deck changes as the game progresses; cards dealt in one round are no longer in the play of the next round. As the dealer deals, the advantages may shift toward or against the casino. So one strategy (card counting) is to mentally record which cards are still in play and which are not and thereby compute the advantage. The smart player bets heavily when the advantage is his or hers, lightly when the advantage is the casino's. Here's an example straight out of Ed Thorp's book. Let's say you've carefully counted cards and believe that precisely two sevens and four eights are left in the unplayed deck; admittedly, a simplistic and rare case, but one that makes the strategy clear. Thorp tells you to place the maximum bet allowed, even if you have to borrow money. He claims that you will certainly win if you unfalteringly stand on the two cards you will be dealt.[13] Your win is guaranteed by mathematics. Luck is not at all involved. The dealer's hand will be either (7,7), (7,8), or (8,8). Either way, his or her total will be less than 17 and, according to the rules of blackjack, must draw. He or she can only draw higher than or equal to 7 and therefore will be forced into a bust.

Thorp was aware of an essay published just a few years before he devised his own strategy.[14] The essay discussed blackjack, as played in Las Vegas casinos, with the objective of finding a strategy to maximize mathematical expectation. Books had been written with the same objective, but "The Optimum Strategy in Blackjack" worked out the necessary sophisticated mathematics to convey a partial solution to the question of when to *draw* and when to *stand*.[15] The strategy is simple. Suppose that D is the numerical value of the dealer's up card, and S is the player's total.

Case 1. (The player's hand is hard—there are no aces): Let $M(D)$ be an integer defined as

$$M(D) = \begin{cases} 13 & D = 2, 3 \\ 12 & D = 4, 5, 6 \\ 17 & D \geq 7, D = (1, 11) \end{cases}.$$

If $D + S < M(D)$, then the player should draw, otherwise stand.

TABLE 10.3
Conditional Probabilities for Outcomes

D	2	3	4	5	6	7	8	9	10	(1,11)
E	.09	.123	.167	.218	.23	.148	.56	−.043	−.176	−.363

Source: Roger R. Baldwin, William E. Cantey, Herbert Maisel, and James P. McDermott, "The Optimum Strategy in Blackjack," *Journal of the American Statistical Association* 51:275 (1956): 429–39.

Case 2. (The player's hand is soft—contains at least one ace): Let the player's total S be the maximum total, counting aces as elevens. Then let $M(D)$ be defined as

$$M(D) = \begin{cases} 18 & D \leq 8, D = (1, 11) \\ 19 & D = 9, 10 \end{cases}.$$

If $D + S < M(D)$, then the player should draw, otherwise stand.

"The Optimum Strategy in Blackjack" claims that if the player uses the optimum strategy as explained in the essay, his or her overall mathematical expectation would be −0.006 with conditional expectation E given in table 10.3.

Thorp knew of this essay and began thinking about devising a winning system that could enhance and extend the "Optimum Strategy" to include all contingencies. After thoroughly studying the game, he used an IBM 704 computer (high speed in the early 1960s) to improve and generalize previously known strategies to compute the player advantages and optimum strategies by removing certain cards from the deck.[16] For example, who gains (player or casino) when four aces are removed from the deck? Twos? Threes? Fours? Who gains when four fives are removed? It turns out that, with all fives removed, the player has an advantage of 3.6 percent and that it does not matter whether the deck is full. For a full strategy, complete with punch-out crib cards to bring to a casino, see Thorp's book.

Craps has very simple rules and many variations. It is a derivative of the ancient game called hazard, which has a debatable history alluding to Arabic origins—the Arabic word *al zar*, sometimes pronounced as *azzah*, means "dice." Street craps is played around the world, where men are known to recklessly bet a week's wages or lose brass watches or coats on throws of a pair of dice with finger snaps and mysterious luck-wooing expressions such as, "Ah, baby! What did the rabbit say? Gi'me luck! Get my gal a pair of new shoes!" or any of many of curious, inexplicable blurts. The excitement is always evident, no matter where or who plays. A 1905 *New York Times* article alleged that the millionaire who plays poker to forget his business worries would not get half the enjoyment from a thousand-dollar jackpot that the poor craps shooter would get from a win of two bits.[17]

Unlike blackjack, craps odds are relatively easy to compute. First let's review how the game is played. There are all sorts of betting possibilities, but these are the basic rules. Players bet by contributing to the pot. One player, designated as the *shooter*, rolls two dice. If 7 or 11 turns up on the first roll he wins the pot. If 2, 3, or 12 turns up on the first roll, the shooter loses the pot. If 4, 5, 6, 8, 9, or 10 turns up on the first roll, that number is marked as the *point*, and the shooter must continue to roll until either the point comes up (in which case he or she wins the pot) or until he or she "craps out" by rolling a 7 (in which case he or she loses).

After the first roll there are three possibilities: the shooter wins, loses, or continues. The probably of winning is computed as follows: There are 6 ways for two dice to give 7, and 2 ways to give 11. So the probability of winning on the first toss is

$$\frac{6}{36} + \frac{2}{36} = \frac{8}{36}.$$

Similar arguments give the probabilities of losing and continuing as 4/36 and 24/36, respectively.

Now let us assume that the player continues after the first roll and tries to make his or her point. Assume also that we know the point and that the probability of rolling that point is p. What is the probability of making that specific known point before rolling a 7? Let us label that probability W_p. On the second toss, the shooter could hit 7, make the point, or have to roll again. So the probability of winning on the second toss is also the probability of making the point before rolling 7. But the probability of rolling 7 on any toss is 6/36 or 1/6. The probability of rolling again is the same as the probability of not rolling 7 and not making the point, which is

$$r = 1 - \left(\frac{1}{6} + p\right).$$

Hence,

$$W_p = p + pr + pr^2 + pr^3 \ldots.$$

This is an infinite sum that can be more simply written as

$$p\left(\frac{1}{1-r}\right).$$ [20]

Since

$$r = 1 - \left(\frac{1}{6} + p\right),$$

the previous expression may be written as simply

$$\frac{6p}{1+6p}.$$

So

$$W_p = \frac{6p}{1+6p}.$$

For example, suppose the shooter's point is 5. Then $p = 4/36$, and the probability of making the specific known point is $2/5 = 0.4$. (Our calculation was based on an infinite series, so you might question whether it assumes an infinite number of tosses. Yes it does, but since r is less than 1, the sum converges, and the sum of the first four or five terms will already be fairly close to the value of the completed infinite sum.)

Since we don't really know the shooter's point in advance, we must assume that it is any one of the numbers between 4 and 10, excluding 7. Thus, we compute the values of p for each of the possible points.

There are 3 ways for 4 to appear: (1,3), (2,2), or (3,1).

$$\text{Hence } p = \frac{3}{36} = \frac{1}{12} \text{ and } W_p = \frac{1}{3}.$$

There are 4 ways for 5 to appear: (1,4), (2,3), (3,2), or (4,1).

$$\text{Hence } p = \frac{4}{36} = \frac{1}{9} \text{ and } W_p = \frac{2}{5}.$$

There are 5 ways for 6 to appear: (1,5), (2,4), (3,3), (4,2), or (5,1).

$$\text{Hence } p = \frac{5}{36} \text{ and } W_p = \frac{5}{11}.$$

There are 5 ways for 8 to appear: (2,6), (3,5), (4,4), (5,3), or (6,2).

$$\text{Hence } p = \frac{5}{36} \text{ and } W_p = \frac{5}{11}.$$

There are 4 ways for 9 to appear: (3,6), (4,5), (5,4), or (6,3).

$$\text{Hence } p = \frac{4}{36} = \frac{1}{9} \text{ and } W_p = \frac{2}{5}.$$

There are 3 ways for 10 to appear: (4,6), (5,5), or (6,4).

$$\text{Hence } p = \frac{3}{36} = \frac{1}{12} \text{ and } W_p = \frac{1}{3}.$$

The probability of winning is, therefore,

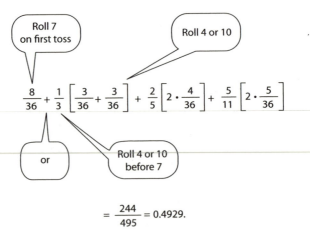

$$= \frac{244}{495} = 0.4929.$$

Though the game is not entirely in favor of the shooter, it is very close to being a fair game.

Aside from lotteries, slot machines have the worst betting values. Walk into any casino and you will find machines advertising to fools. "This bank pays 99 percent" is what the blinking neon sign over the slot machine says. Of course it does, but all it means is that over an infinite number of plays of the slot machine the player will receive 99 percent of all the money he or she already spent. The hopeful, subliminally misleading interpretation is that he or she is almost certain to win. The trusting player, *hot* with a fistful of coins, is sure to be taken in. In 2005 Americans spent more than $33 billion on slot machines in casinos. That figure does not include revenues from other venues.

Slot machines used to be mechanical. A lever would start three, four, or five reels spinning. On each reel were a specific number of symbols (generally 20 or 22, including blanks) with one jackpot symbol. If the three reels line up three jackpot symbols, the jackpot is won, though with 20 symbols there are 8,000 distinct combinations, only one of which is the jackpot combination. No skill involved, though some believers feel some kind of control by the way the lever was pulled; others hold wishful strategies. The odds are simple to calculate: if each reel has N symbols, then, on a three-reel slot machine, the probability of hitting the jackpot is

$$\frac{1}{N} \cdot \frac{1}{N} \cdot \frac{1}{N} = \frac{1}{N^3}.$$

So the odds are $N^3 - 1$ to 1. That is, there are $N^3 - 1$ ways to lose and 1 way to win. If $N = 20$, as in our case, then the odds are 7,999 to 1.

A fair expected payout for a $1.00 slot would be 7,999 to 1. In that case, over the long run a player would break even. Surely the house would not allow that, so the payout is less, perhaps 6,000 to 1. You would have to play, on average, 8,000 times (spending $8,000) to receive $6,000. The expected value is −0.2. With five reels the odds are much worse; the probability is

$$\frac{1}{20^5} = \frac{1}{3,200,000},$$

odds of 3,199,999 to 1.

Modern slot machines—the kind that began to operate in the late 1980s—are entirely electronic, computerized, and programmed to display a line of symbols entirely determined by a random number generator. They are not limited to three to five reels or ten to twenty symbols and could have as many as a hundred symbols and ten reels, but they are programmed in such a way that the odds are in favor of the house. Unlike the mechanical slot machine, the electronic slot is programmed in such a way that each symbol has a different chance of occurring. Players may see reels spinning, but they are just for show; they are either virtual reels on a video screen or mechanical reels programmed to stop at the predetermined symbol. Though there may be ten different symbols, the chances of a strawberry appearing may not be 10 to 1 but programmed to appear, say, once in fifteen spins. It works this way: The computer is simply selecting random numbers, say from 1 to 10,000, one random number for each reel. Each distinct symbol (on a list of, say, 100) is assigned to several different numbers. So, perhaps a cherry is assigned to both 1 and 81 and a strawberry is assigned to 5, 25, and 8,005. With 100 symbols and 10,000 numbers, we might expect that every symbol is associated with 100 different numbers. Then, for a machine with five reels, the probability of winning would be 0.0000000001. However, that is not the case, for the machine is programmed so that the jackpot symbol is associated with, say, just 50 different numbers and therefore comes up more rarely.

This makes it difficult for the outsider to calculate the odds of winning the jackpot. The payout percentage ranges from 92 percent to 98 percent, depending on the location, type of machine, and the amount it costs for a single play. To figure this out, consider the simple case of a machine with 1 reel and 10 symbols labeled by the numbers 1 through 10, and with payouts listed in table 10.4.

The house edge on table games such as roulette, blackjack, and craps is normally expressed as a percentage. For games like poker

TABLE 10.4
Payouts on a Slot Machine with
One Reel and Ten Symbols

Symbol	Payoff
1	$4
2	0
3	$2
4	0
5	$2
6	0
7	$1
8	0
9	$.50
10	0

and blackjack, that percentage assumes that the player has played with perfect basic strategy. However, the house edge for slots is different. It is measured in terms of what is called *payout percentage*, which is computed as percentage of the amount of money paid out in the long run compared to the amount played. A machine with a 95 percent payout percentage would return (in the long run) $95 for every $100 played. According to *Strictly Slots* magazine, the general payout percentage in 2006 was roughly between 92 and 98 percent.

Now suppose that each time you play the reel stops at the next number on the table, starting with the first. What's the payback? You paid $10 to play and received $9.50. So your payback is 95 cents on a dollar. A sign on this machine could then legitimately say, "This bank pays 95%." It is possible, though very improbable, that each time you play, the machine randomly picks the next number and cycles through the numbers 1 through 10. Improbable, yes, but the reasoning here is analogous to what happens when a single die is tossed: each of its six faces should appear equally often in the long run, and so, in six tosses each of its six faces should appear approximately once—in the long run. Hence, we may think of slot machine plays as we think of dice tossing: each of its ten symbols should appear approximately once, in the long run. Nevertheless, the justification for the machine's misleading advertisement is correct. Moreover, the computation analysis of a machine's payout percentage, using this cycling argument, is the same for all slot machines, even computerized machines with virtual reels and millions of possible combinations on imaginary reels.[18]

Let's say there are 100 symbols (including blanks) and we want a payout of 95 percent. We calculate that a cycle at 1 coin per play would cost 100 coins. Ninety-five percent of 100 coins is 95 coins. (The value of a coin does not matter.) In that case the machine should be paying back 95 coins for a completed cycle. It would assign a large share of those 95 coins to the jackpot symbol and fewer coins to the other symbols. Adding reels only makes the jackpot larger, but the computational procedure is similar, though more complicated— comparable to adding dice to a game played with a single die. Payout percentages for particular machines are difficult to compute; the casino operators consider them classified information. To make those computations you would have to play an enormous number of times, and even then, your computations would be compromised by some of the bigger payouts that rarely happen.

It has been suggested that casinos should advertise a number L to give the expected loss per hour. Let B be the average number of bets per hour, M be the amount bet per hour, and E be the house edge. Then $L = B \cdot M \cdot E$ would represent the amount the player loses per hour over the long run. By computing L, the player would know his or her entertainment cost and be in a position to judge whether or not he or she could afford to continue playing. Casino managers know the value of L, but they would be fools to display it.

Sometime in the late 1990s, casino managers discovered that the old mechanical spinning reel machines, those one-armed bandits of the 1970s, encouraged more excitement than the video display machines. Manufacturers went to work designing machines that could renew that excitement and so had the random number generator simply pick numbers and direct real reels to stop at the symbol (or blank) determined by the randomly picked number. Players may now watch real reels spin, thinking that the reels will mechanically

FIGURE 10.1. The Watling "Bird of Paradise" slot machine, an example of the one-armed bandit. Photo courtesy of www.slotsetc.com.

stop at Fortune's will. In reality, those reels will stop on a symbol predetermined by the random number. Their artificial stops are obvious, and yet the excited slot gambler gripped in the emotional moment believes what he or she wants to—or needs to—see.

The mechanical arm of a true one-armed bandit of the pre-1970s played an important psychological role. It provided an illusion of control, the bait on the recruiting hook that turns rookies into persistent gamblers. You pulled that solid arm with a determined feeling that a managed speed and firm grip could influence the result. The clicking sound of the completely unnecessary internal ratchet contributed to the impression that gears actually connected the turning arm to spinning reels.

And now the one-armed bandit is making a comeback. Casino managers and game designers learned that electronic machines diminish the player's illusionary sense of control, so the lever is back with its impressive imitation of the real thing.

Under the misguided belief in commanding control, the one-armed bandit gambler abandons any sense of risk, disregards seriously unfavorable mathematical odds, and is ensnared in a net of complex emotional powers that potentially lead to harmful addictive behaviors.

That feeling of control is the stimulant of any gambling game. The craps shooter generally believes he or she can influence the dice by particular throwing techniques; the roulette player thinks that he or she knows the winning bet the moment the chips are placed, and often trusts that that knowledge will influence the result; the blackjack player impressionistically anticipates the value of the next card dealt. Though the gambler's losses should conclusively cast doubt on controlling outcomes, they seem to have no rational effect on his or her belief in having powers over chance. The question is, why don't they?

The answer is not simple. The psychology of gambling behavior is indefinite, deep, and young. Psychologists have been earnestly studying gambling behavior for more than a century, critically modifying a wealth of speculative theories based on analyses of relatively few clinical trials. Some hypotheses are accepted, but many remain speculative.

Part III

THE ANALYSIS

Gambling behavior and addiction have been studied since the turn of the twentieth century when psychopathologists began investigating repressed libidinal impulses, the conflicts between the id and superego. The tensions between the demands of conscience and the performances of the ego led to fascinating theories of gambling addiction. Some of the early theories have been disputed and some still claim that the impulsive gambler may not have a conscious will to win but, rather surprisingly, an internal conflict causing an unconscious desire to lose.

Though the pathology of compulsive gambling is still not satisfactorily explained by any one theory, it looks as though, in certain personalities, dormant gambling tendency can be awakened by a variety of environmental factors that play a fundamental role in determining motivation for gambling and other addictions.

Gambling impulses are not easily categorized. Certain gamblers are driven by unrealistic desires to control fortune through some weaving of skill, greed, and luck. They have illusions of influence over randomness and are compelled to impose some sort of order or meaning on their future.

Selected inherited histories of neurotic tensions stored in the id, possibly guilt, may be responsible for the urge. Addictions could be genetically determined or subliminally dependent on life's experiences. They may come from a driving need for stimulation, an antidote to boredom perhaps caused by overactive dopamine nerves.

♣ CHAPTER 11 ♣

Let It Ride

The House Money Effect

I hold the Fates bound fast in iron Chains
And with my hand turn Fortune's wheel about, . . .
—*Christopher Marlowe, Tamburlaine*

I was eight years old when I lost my lucky aggie shooter marble to Jerry Cutler playing immies. It was streaked with colors. Jerry was one of the older kids. He acted as the owner of a gambling establishment, a rake; his turf was the walls and sidewalks outside our six-story apartment building in the Bronx. On summer afternoons the street smelled of chestnuts and peanuts from a pushcart vendor. A noisy street. If the rag and junk wagon were not clanging pots and hubcaps together, a knife sharpener might be flying sparks at screeching high frequencies over the clamor of unending street games active in uninterrupted concert. The corner of Harrison and Tremont was alive night and day, neighbors chatting about the good old days that seemed to be few but memorable, the hard old days when everything was a challenge, the war days, Depression days, those days—"not like now when children have everything so good"—and these days with the outrageous price of gasoline at twenty-one cents a gallon and the price of a gallon of milk for just about the same amount. These regular symposia happened during bridge games on unsteady card tables or while small groups of mothers met to knit sweaters for the next winter six months away.

The games had variations. To start, each player placed one marble inside a square drawn on the sidewalk. Players took turns shooting marbles from some set distance with the flick of a thumb against an index finger to knock as many marbles as possible out of the square. A player got to keep any marbles he knocked out. Jerry's take was different. He had a cigar box with three holes cut out, each slightly bigger than the next. If a player shot his marble into any of the holes he would win one, three, or five of Jerry's marbles, according to the size of the hole. Any marbles that didn't enter the box through one of the holes was Jerry's to keep. The odds were always on Jerry's side and he bankrolled the house as if he were Diamond Jim Brady.

We didn't only gamble for marbles. We also pitched pennies against a wall—the closest to the wall won all the other pennies of the round. We flipped and traded baseball cards, and sometimes spun playing cards for pennies. But the loss of my aggie was a heart-felt blow. My big brother gave it to me along with a Phil Rizzuto card and a 1939 New York World's Fair three-cent special issue postage stamp he once traded with Jerry Cutler for a piece of loadstone. So I had to win that aggie back. Jerry offered me a chance. Get a shooter through the smallest opening—about the size of my special issue postage stamp—and I could have my aggie back. I shot and lost another less favored marble. I shot again and again and came closer and closer. One went in and bounced back out.

"That doesn't count!" Jerry shouted.

"Why not? The rule says you just have to get it in, not that it has to stay in!" I argued.

But Jerry was much bigger and in the Bronx, bigness earned respect.

Those close shots became the teases for more challenges to recoup my losses. I was down to my last two shooters after losing bags full. They were my dullest marbles. With each shot, I thought I had luck behind me. And then it happened. With a crowd of kids watching from the sides, yelling and screaming like spectators at a cockfight, my next-to-last shooter disappeared through the small hole in Jerry Cutler's box. All I got back was my color streaked aggie shooter.

Now, I was a normal kid, with a normal feeling that something as special as that aggie given to me by my big brother must be a lucky thing, a thing to bring me luck. So what was I to do but challenge Jerry to get my sack of marbles back and break even? It was a lesson in luck, for not only did I lose, but I had to trade my Phil Rizzuto card for five more plain shooters and then my 1939 New York World's Fair three-cent special issue postage stamp. By the time it was all over, I was behind by fifteen marbles and had lost my other treasures. Quite a lesson for an eight-year-old!

You would think I'd have learned a lesson. No! Soon I felt lucky again and started gambling again, this time for money. The stakes were small, just pennies and nickels, but I wasn't playing for the value of money. Rather, it was for the testing of luck—my luck. I felt— through an eight-year-old boy's kind of thinking—that I had a certain power over the odds, a certain control that I couldn't explain, except to tell myself that if I willed an event to happen, it would by a kind of voodoo control over nature. That feeling would not leave me until much later in life when the boy in me grew up to learn that voodoo contradicts too much of what we know about science.

And yet . . . .

Knowing all that I know, at times I still resort to a sort of reverse voodoo when I'm anxious about some danger that may involve a loved one miles away. I try hard to hold thoughts to prevent any harm. As the watched pot never boils, so the thought of bad events will never happen. Thinking of winning the lottery seems to make that unlikely event even more unlikely.

But shooting marbles is not purely luck. It takes coordination, not any understanding of your opponents the way poker does, but there is another kind of skill involved that uses the force, speed, and spin you put on the marble.

In my boyhood years it seemed that everyone in my neighborhood gambled at something. My father and friends' fathers spent Saturday afternoons at Yonkers Raceway betting on trotters. We kids weren't supposed to know that, and we surely were not supposed to know that fathers bet part of their weekly pay on numbers. But it was hard to avoid the talk on street corners where bookies stood conspicuously receiving envelopes from passersby. Winning at numbers was purely a matter of luck, though many players would deny that. It seemed that players labored at convincing themselves that there was skill at picking numbers.

So, of course, we kids were smart enough to know that gambling was not a privilege reserved for grownups. We couldn't play the numbers or horses, but by the time I was a teenager, gin rummy was the sport of the neighborhood and money in the form of small coins was rapidly moving between players, and being collected by just one or two players. Some seemed to think that luck was responsible for their success. "He has all the luck!" was the refrain heard at the end of every card game played by Tony Luce—Lucky Luce, we called him. Some said he cheated, others that he had some spiritual gift. The gift was right, but it wasn't spiritual. He played an intense game and knew how to play it, not just with cards but also with humans and human faces. Rummy, poker, and blackjack were played to blasting music so betting could remain covert and bigger pots of money could exchange hands. After a while nobody would play with Lucky, so he would move on, block to block, to places where his reputation was clean, to where he could muster a blackjack or poker game before being suspected of cheating or hustling. Not everyone was naive and not every neighborhood was as accepting as ours, so he would often return grateful to get away with just a black eye or nosebleed.

We first started playing on a stoop in front of Lucky's house directly in front of his mother, who sat on a folding chair forever peeling potatoes, never saying a word to anyone but Lucky. At the time, we all thought that Lucky was—well, simply lucky, even though we were suspicious that it was not entirely luck. I know now that playing gin rummy (a card game whose name suggests its original stakes) can be almost entirely skill—*almost*, because the player still has to have some luck in getting a lucky hand. I know now that a good player could know exactly what cards are in his opponent's hand without cheating.

Mathematics of dealing from a deck is just one element of a card game. Memory, psychological deception, courage, money management, and analysis of an opponent are five additional essentials. To appreciate the extremes of card-playing skill, take the case of the Stu Ungar story. Stu was a professional, one of the best. As in all poker-like games, the object is to acquire a hand of highest value. In reality, Stu was one of the greatest poker players. The following dialogue is from the movie version of the story of his career. He is at a gin rummy game with casino mogul Leo the Jap.

> DEALER: If Stu wins this next game that will be about $8,300 total, Mr. Leo.
> LEO: Nice run of luck.
> STU: Oh, it ain't luck.
> LEO: What then?
> STU: I know you.
> LEO: (*Snickering*) You know me?
> STU: Yea, so I know what you got.
> LEO: Ah, any good gin player learns that.
> STU: Yea, but I know now.[1]

The dialogue comes from a movie script, but it is an excellent portrait of the skills of a master gin player. It's also likely that the actual Stu Ungar had such a skill. A good poker or gin player knows not only the cards but also his or her opponent. A few games are played just to study moves, expressions, habits, hesitations, clothes worn, where drawn cards are placed in the hand, and all those imperceptible, unremarkable movements that to novice card players would be invisible.

> LEO: After four discards? Impossible!
> STU: Double or nothing says I know.
> LEO: That's a bet.

Real gamblers take risks to place bets within bets. Real gamblers know that their wagers are mixtures of knowledge and luck. They also know that their ratio of luck to skill is usually quite small, something the recreational gambler doesn't get. Stu knows nine cards in Leo's ten-card hand—he's not lucky-guessing, but knows for sure and isn't cheating.

> LEO: So ah . . . talk to me, little master.
> STU: Well . . . I know you have a pretty decent hand cause ya always move forward when the cards look good to ya on the deal. Now

that means ya got two spreads cause if ya had three when I hit you with the king of diamonds ya would have knocked . . . which ya didn't. Since I got the queen of diamonds . . . ya got kings.

LEO: (*Shows three kings.*) Brilliant!

STU: Oh, not really. See, but do ya know you always keep your low spreads on the right side of your hand . . . yea, ya do. But with aces, for some reason, ya start on the left until for some reason ya move them over. Now that's interesting, right? (*Leo shows three aces.*) And, I also know ya really, really hate to throw middle cards early, except if you're protected and fishing. So I would bet that seven of hearts came off a pair of them. Since I got the six of hearts, ya got the eight of hearts. . . . (*Leo shows the eight of hearts.*) Along with the eight of clubs . . . (*Leo shows the eight of clubs.*) The six of clubs . . . (*Leo shows the six of clubs.*) And the seven . . . let's see . . . it could be spades or diamonds—I'm not really sure, not that it would matter.

Now Leo is also a brilliant poker strategist who can find his opponent's weak points. He knows Stu. Thinking fast, he offers Stu an alternate bet, a bet he knows Stu would not refuse.

LEO: All right, all right, kid you made your point. How about we just call it an even ten grand and be done with it?

At this point Stu could have walked away from the table with ten thousand dollars. But greed grabbed him and he felt sure that he knew that Leo had a seven of diamonds and wanted to complete the original bet of doubling the $8,300. This is another point where Stu thinks *he knows* Leo—in reality, he is guessing, using luck. He's not thinking, *I'm feeling lucky so I'll take a guess.* After all, he's narrowed it down to even odds that he's right. It's either spades or diamonds. He could have flipped a coin, the kind of thing the amateur would have done. No! He knows it's a coin flip, but he has more information up his sleeve. He needs to know the bias in the coin flip. He hesitates. "I'm not really sure," he says while trusting his keen character judgment skills to determine the bias.

STU: No. Nah, I don't think so Mr. Leo. You're kind of a diamond guy with your suit and ah rings and stuff. So I'm gonna say

diamonds . . . yea, seven of diamonds. Now personally I would
have kept the 7, 7-8 nutcracker together and risked the six . . .
but that's just me.

LEO: I really admire your courage kid, but this makes us even.
(*Shows seven of spades.*)

One may think this is fiction. We know that rummy requires the
usual skill of keen surveillance of cards in play. But how could someone know his opponent's cards simply by observing characteristic
behavior? Perhaps one card could be determined—but *two*? *Three*?
Four? *Five*? Stu Ungar was phenomenally gifted. But we should not
think the skill impossible. Inexplicably gifted humans are capable of
performing the most astounding feats. Just watch Serena Williams
play tennis, Francis and Lottie Brunn speed-juggle, or Philippe Petit
walk the high-wire, or listen to Van Cliburn play Tchaikovsky. Stu
Ungar also had a gift.

Despite winning millions during his poker career, Ungar died at
age forty-five in a Las Vegas motel room on November 22, 1998, with
no assets to his name.

In this dialogue we see a mysterious mixture of skill, greed, and
luck—all woven into an illusionary control of destiny. And it is that
kind of control that drives the forces of the game. It is the same
control that comes from the feeling that luck is a matter of defeating
destiny, that games of pure chance are games of out-and-out skill.
Such control was well understood in much of Dostoyevsky's writing
but it was most particularly relevant in *The Gambler* where Antonida
Vassilievna Tarassevitcha, the Grandmother, a rich, seventy-five-year-
old invalid and grande dame of Moscow, comes to Roulettenberg, a
fictitious German spa town with a casino, and makes a commotional
grand entrance into the roulette-salon where she is wheeled to the
roulette table. She plays with her peculiar notion of luck by repeatedly giving Alexis Ivanovitch, the narrator, ten gulden coins to stake
on zero. When he tells her that the number of chances against zero is
thirty-six and that zero had turned up just moments ago, she retorts,
"Rubbish! Stake please."[2]

Alexis Ivanovitch is the tutor in the entourage party of the Russian
general Zagoryansky. The entire story takes place in the mythical

town of Roulettenberg, where we find some traces of reminiscent reflections of Dostoyevsky's own gambling escapades at the tables of Wiesbaden with the beautiful Polina Suslova and borrowed money from the Fund for Needy Authors.[3]

Alexis Ivanovitch warns her that zero is not likely to turn up again that night. To this, she insists, "Rubbish, rubbish! Who fears the wolf should never enter the forest. What? We have lost? Then stake again." A second and then a third stake of ten gulden were lost. By that time the Grandmother could not sit still. And when the fourth stake was lost, she was fuming at the croupier, "To listen to him! When will that accursed zero ever turn up? I cannot breathe until I see it. I believe that infernal croupier is PURPOSELY keeping it from turning up." This time she gives Alexis two gold pieces (twenty gulden). When he protests, she exclaims, "Stake, stake! It is not YOUR money." So the two gold pieces are bet. The ball goes spinning round the wheel before settling into one of the notches. The Grandmother sits petrified, with Alexis Ivanovitch's hand rigidly clasped in hers.

"Zero!" calls the croupier.

Now I have to tell you about my dream. It is a story of control similar to that of the Grandmother in *The Gambler*. In the dream I win eight million dollars on a promotional quiz game after answering the question, "Who is the author of a book published in 1900 called *The Interpretation of Dreams?*"

In the excitement, I half wake from that dream into a dream one level up toward real consciousness. Very disappointed that I actually won nothing, I withdrew a hundred thousand dollars from my bank account and hopped on a Greyhound bus to Atlantic City where I took a seat at a roulette table. I put $5,000 on zero (supposedly the house limit) and calmly watched the croupier spin. The ball spun smoothly, whirring, around the rim and landed in notch 27. With another $5,000 on zero I sat, eased by the thought that I could continue betting on zero another eighteen times. The croupier called out 6. I staked another $5,000 on zero, rightly thinking *any number is as good as any other, so why not leave my chips on zero?* Again I lost and placed another bet on zero. I again lost. But on the seventh stake, the ball seemed to hum more silently around the rim before spiraling into a pocket and I thought I saw that it had fallen into the zero pocket, though I could not be sure before the wheel slowed down.

"Zero!" cried the croupier with a glance toward me.

I was ahead by $40,000 and should have quit at that moment, *but*, I think, *the $40,000 is extra money to play with, so what the hell!* A crowd gathers as I put my entire winnings on zero once again. (Somehow the dream forgot that the house limit was $5,000.)

The croupier gave me a quick glance, probably thinking, *what a fool.* The wheel was spun and ball tossed to whirr, as it always does, around the smooth rim before it finally came to rest in one random pocket of the silent spinning wheel.

"Zero!" called the croupier in astonishment.

And behold, the lucky man in this gambling sea came up with a $1.4 million fish in his mouth.

Now, let's take a momentary break from this story to think about what the average person would do at this moment. One choice would be to just walk away from the wheel. But wait: Why not play with some small part of the house money? Say, $400,000? This is falling into a gambling trap called the *house money effect*, the impression that money in the pocket given away by the house is free to be gambled at high risk. Or, perhaps more cautiously, $40,000? After all, the chances of the ball falling on zero again are the same as they always were—1 in 38. What can be done once, can be repeated. And that's exactly what my dreaming sleep brain thinks. *I can make thirty-five $40,000 stakes with the house's money, so surely I have a reasonable chance.*

But this dreamer made stake after stake of $40,000 all night. Croupiers changed guard and by the thirty-fifth wager I had wiped out all my gains. My next bet would have to be with my own money, and being accustomed to $40,000 stakes, just three would wipe out the $100,000 I had when I arrived in Atlantic City. But a real gambler keeps a history of his winnings in mind, and I could not forget that just hours earlier I had $1.4 million in my pocket. I had to retrieve what I once had, even though what I once had was house money. So, again, I staked $40,000 on zero.

"Fifteen!" the croupier called, careful to avoid a noticeable glance at my eyes.

And once again, I put down $40,000 worth of chips.

"Twenty-one, red!" yelled the croupier.

I had lost $80,000 of my money and was entrapped by a *break-even effect*. So I continued to gamble and lose until I was so much in debt that I could no longer retrieve my losses without borrowing. But borrowing put me more in debt to my lenders. We see how things can turn so quickly. I would forever chase my losses. Fortunately for me it was all just a nightmare. But not everyone is so fortunate to learn so much from a dream.

Chasing losses to break even is the gambler's peril. Studies show that novice gamblers track their account balances mentally with a tendency to update winnings and not losses.[4] Excessive gamblers tend to recall long strings of losses ending in wins and thereby disregard their losses as means to ends. Frequent and prolonged gambling inevitably involves frequent losses, losses that generally outweigh gains by a sizable margin. But the losses are forgotten and the big gains not. The mind tends to discount losses in the face of possible wins. In everyday life such discounting may be necessary for survival, or at least protection against discouragement. Biased memories adhere to the illusion of control to rear the kind of behavior that favors omnipotence, the feeling that luck is a matter of beating chance, the imprudent belief that even games of pure chance, such as slot machines, roulette, and lotteries, are games of skill.[5]

That fantasy of controlling chance—the overconfident belief in one's personal luck—is the gambler's illusion. It is the daring that confuses chance with skill. The psychologist Ellen Langer experimented with the notion of confidence in gambling games. According to her, even in games of pure chance and no skill, the well-dressed gambler feels more confident and wagers more aggressively against disheveled opponents than against the better groomed. Langer's experiments involved a game of simply drawing cards from a well-shuffled deck—highest hand won. The more dapper gambler took riskier chances under an illusion of being better at controlling chance than his or her opponents. Such exaggerated confidence does seem to increase the probability of winning when the game involves skill; for example, in poker, the self-assured bluffer is better at deception than the timid bluffer who reveals him- or herself through uneven performance.

Casino corporations have researched gambling preferences and habits to learn which games, machines, and environments encourage further play. They have tweaked the environment to influence everything from the number of gambles per minute to jackpot sizes in order to affect the illusion of skill and arouse moods of impending good fortune to entice further gambling.[6] These corporations understand that the attraction is the rapid payout, the broad range of odds, and the high degree of emotional involvement. One wonders why they don't just give away a few coins to hook punters.

♣ CHAPTER 12 ♣

Knowing When to Quit

Psychomanaging Risk

> Some people are luckier than others,
> whereas chance is the same for everyone.
> Neither luck nor chance can be directly affected
> by people, but one can take certain actions in order
> to take advantage of good luck and avoid bad luck.
> —*From a survey of blackjack players at an Amsterdam casino*

The beautiful sylph Gwendolen in George Eliot's *Daniel Deronda* is caught in the excitement of playing roulette when we find her daintily gloved, adjusting her winning coins for a moment before pushing them back to bet again with resolute choice and belief in luck as a possession. As she began to believe in her own luck, others too began to believe in it. She envisions herself as a goddess of luck with a worshiping entourage that would watch her play.[1] Fortune and bets continue until her wandering eyes first meet Deronda's measuring gaze, when something happens to uproot her inner conflict; at that moment her uncommitted belief in the evil eye sways to submission. She loses her luck and her stake. "Faites votre jeu, mesdames et messieurs," calls destiny as Gwendolen's arm reaches to dump her last short stack of napoleons. "Le jeu ne va plus," says destiny.

"Was she beautiful or not beautiful? . . . She who raised these questions in Daniel Deronda's mind was occupied in gambling."[2] So begins George Elliot's Victorian society novel. Rarely does one come across such an alluring first line of only six words. Gwendolen is beautiful to be sure, but spoiled and flirtatious, and—because of family financial ruin—is faced with the unhappy prospect of gambling with the stake in winning or losing her own happiness by marrying the depraved, cruel, controlling, but rich Grandcourt for money.

Gwendolen certainly had beauty, a manner of firm choice and a belief in her own luck. Indeed, those were her true properties, but she did not possess luck in the same way she owned her beauty. Her manner of firm choice gave her the belief she could control her own luck, but she was not aware—as nobody is—that luck, when too strongly assumed, is often escorted by greed.

Consider this experiment: You are shown two unmarked suitcases; one contains $1,000, the other $200,000. You are given options of either collecting $101,000 (roughly the average between the two extreme amounts) or keeping one of the unmarked suitcases. Which option do you take? On December 10, 2007, NBC broadcast its hit show *Deal or No Deal* in which Shalanda London, a contestant from Cedar Hill, Texas, was given, essentially, this same deal. She could have walked home with a secure $101,000 (before taxes). But, with her adamant feeling that luck was on her side, she turned down the deal and ended up with the suitcase containing $1,000. It probably cost her more than $1,000 in clothes and travel to appear on the show.

Before making her decision about which suitcase to pick, her mind had to do something analogous to flipping a coin. Since she did not know which suitcase contained the $200,000, she had to guess. Assuming that each suitcase was identical with nothing to favor one choice over the other, the odds of mentally picking one over the other must have been even. Now it's true that she came with nothing and therefore would lose nothing, except travel expenses and, perhaps, new clothes for the trip to LA. But let's think of it this way: She was being told by the program's host, Howie Mandel, though not exactly in these words, "Here is $101,000. Free. Payment for you to stop playing the game now! But, for that $101,000, I'll offer you the opportunity to almost double it by flipping a coin. If the coin comes up heads, you almost double your money. If it comes up tails, you win only $1,000."

Why wouldn't she simply take the free gift of $101,000, a windfall of profit from a game that predicts even odds of winning and losing? There are two possible answers, each calling for more study. The first is simple greed: $200,000 is *better* (better in the sense of more) than $101,000. But the second is far more multifarious. Could Shalanda have been extending and preserving her cherished few moments of fame before a live audience of a few hundred supporters and tens of millions of television viewers? This is the performance sensation, by which a person would tell innermost secrets in front of twenty million people about something he or she would never divulge to his or her closest friend.

Dostoyevsky was keenly aware of the fame effect, for the Grandmother fell victim to it after winning eight thousand rubles (almost $300,000 in 2009 dollars) through her luck of an ivory ball falling into the zero pocket of a roulette wheel *three times in succession*. She had the sense to say, "Enough, let's go home. Wheel my chair away." But let's remember that there was a crowd around her and, as her chair was wheeled out of the gaming salon, everyone wanted to congratulate her. People were pointing to and talking about the Grandmother. Many moved toward her chair to get a better view. Ladies kept a curious gaze, staring directly at the Grandmother.[3] And what did she do at four o'clock the next day? Set out for the casino for more roulette. Those were moments of glory for the arrogant cantankerous old lady who, before her audience, became merry like a child in play.

At the time this book was written, no contestant of *Deal or No Deal* had ever won the top prize of a million dollars. The game is a model of risky choice theory, greed, compulsivity, fame craving, ignorance of expected value, and belief in the possession of luck. It may be only a game, but it is also an experiment in psychology and economics combined.

Understandably, it turns out that there is not much funding for experiments designed to give money to a large sample of potential gamblers just to study risk behavior. So game shows have become the experimental platforms by which economists can test some old thought experiments that have been around since the 1950s.

TABLE 12.1
Prize Amounts in *Deal or No Deal*

$.01	$1,000
$1	$5,000
$5	$10,000
$10	$25,000
$25	$50,000
$50	$75,000
$75	$100,000
$100	$200,000
$200	$300,000
$300	$400,000
$400	$500,000
$500	$750,000
$750	$1,000,000

The show originated in the Netherlands as *Miljoenenjacht* (Chasing millions) in 2002 and is now broadcast in more than forty-six countries. The game starts out with 26 closed suitcases, marked 1 to 26, brought on stage by 26 scantily clad models. The contestant knows that 26 monetary prizes are randomly distributed among the suitcases and can see the prize amounts displayed on a board (see table 12.1).

She is asked to select one suitcase to keep, but not open until the end of the game. Let's call that case *X*. There are nine rounds. In the first she is asked to select six cases to open. The monetary values in those six cases tell her what is not in case *X*. It's not a memory game—the excluded cases are marked on a large electronic display board for all to see—though, perhaps, a memory version of the game would add excitement. Next the banker offers to buy case *X* for a relatively small amount (at times calculated to be as low as 10 percent of the expected value of case *X*). If the contestant says "Deal," the game ends and she goes home with the banker's offer. If she says "No Deal," the game continues to round 2, when she must open five more cases before the banker presents her with a new offer for case *X*. The number of cases opened in each round is displayed in table 12.2.

TABLE 12.2
Number of Cases Opened in Each Round

Round	1	2	3	4	5	6	7	8	9
Number of cases to open	6	5	4	3	2	1	1	1	1

The game continues until either the contestant accepts the banker's offer or the end of the ninth round when there are just two cases left—case X and one other case in the hands of the last standing model. If the banker's offer is refused after the ninth round, that last case is opened and the contestant goes home with case X. Some variations of this game include swapping an unopened case with case X. Such a decision brings other complex dilemmas that would take us in another direction.

Surprisingly (or perhaps not surprisingly) the biggest losers and the biggest winners have something in common. According to recent studies conducted by a team of Dutch economists, both the biggest losers and biggest winners have a low degree of risk aversion.[4] These are risk seekers who reject bank offers that exceed expected values, those mathematically calculated amounts likely to be in case X.

You may suggest that this is not a valid economic experiment in gambling; after all, the contestant was given free money to play with. But it's no different than giving a young boy a quarter to bet on a horse he knows nothing about. Money given away by the house is money in the pocket, considered free to be gambled at high risk. In a landmark essay, the American behavioral economists Richard Thaler and Eric Johnson wrote about a finding they called the *house effect*—under some circumstances, an earlier gain can increase a subject's eagerness to gamble and an earlier loss can decrease his or her willingness to take risks.[5]

Take the case of Heather McKee, an Oklahoman pig farmer who, in playing *Deal or No Deal*, turned down an offer of $207,000 and ended up with one penny.[6] At the end of each round she did not take the deal of the bank's offer. Table 12.3 shows the distribution

TABLE 12.3
Heather McKee's Nine Rounds on *Deal or No Deal*, January 3, 2008

Round 0

Heather McKee chooses Case #22 to keep.

Round 1

Cases chosen	2, 15, 6, 24, 8, 19
Unopened cases	1, 3, 4, 5, 7, 9, 10, 11, 12, 13, 14, 16, 17, 18, 20, 21, 22, 23, 25, 26
Bank offer (BO)	$63,000
Expected value (EV)	$152,130
BO as percentage of EV	41%

Round 2

Cases chosen	13, 21, 25, 1, 11
Unopened cases	3, 4, 5, 7, 9, 10, 12, 14, 16, 17, 18, 20, 22, 23, 26
Bank offer	$82,000
Expected value	$169,454
BO as percentage of EV	48%

Round 3

Cases chosen	4, 16, 7, 10
Unopened cases	3, 5, 9, 12, 14, 17, 18, 20, 22, 23, 26
Bank offer	$148,000
Expected value	$226,437
BO as percentage of EV	65%

Round 4

Cases chosen	14, 9, 5
Unopened cases	3, 12, 17, 18, 20, 22, 23, 26
Bank offer	$207,000
Expected value	$301,939
BO as percentage of EV	68%

Round 5

Cases chosen	20, 23
Unopened cases	3, 12, 17, 18, 22, 26
Bank offer	$45,000
Expected value	$69,251
BO as percentage of EV	65%

(*continued*)

TABLE 12.3
(*continued*)

Round 6	
Case chosen	18
Unopened cases	3, 12, 17, 20, 22, 23, 26
Bank offer	$71,000
Expected value	$83,002
BO as percentage of EV	85%
Round 7	
Case chosen	3
Unopened cases	12, 17, 20, 22, 23, 26
Bank offer	$2,500
Expected value	$3,752
BO as percentage of EV	67%
Round 8	
Case chosen	26
Unopened cases	12, 17, 20, 22, 23
Bank Offer	$4,000
Expected value	$4,986
BO as percentage of EV	80%
Round 9	
Case chosen	12
Unopened cases	17, 20, 22, 23
Bank offer	$5,500
Expected value	$4,980
BO as percentage of EV	110%

table indicating the values in each case. Heather's chosen case was 22 (containing $0.01).

We see that in each round, except the last, the bank offer was below the expected outcome. In the beginning the difference was great, but there was a convergence toward the end. That is how the producers keep the game going and the entertainment factor high.

In a show aired on February 4, 2008, the suitcases contained ten $1,000,000 prizes in the same twenty-six cases.[7] To the novice, it might seem as if the chances of winning one million dollars are increased.

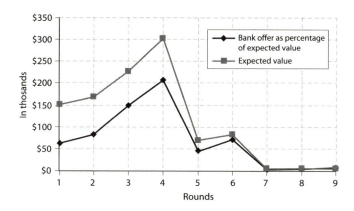

FIGURE 12.1. Difference between bank offer and expected value.

Shouldn't they be? Well, yes, there is a better chance that the one case chosen in the beginning contains $1 million, but to get it means having the guts to continue the game all the way to the end, passing all the bank offers, which will inevitably go down in the later half of the show because there is a greater chance of eliminating million-dollar cases. The point to think about is this: $1 million is the maximum, so with every million knocked out in any round, the bank offer will go down. Since there are more $1 million cases in play, there are more million-dollar cases ready to be eliminated. The contestant, Becky Matheny of

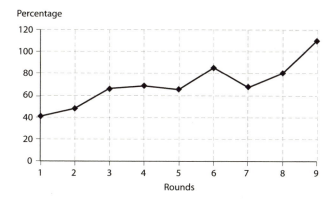

FIGURE 12.2. Bank offer as percentage of expected value.

Mesa, Arizona, clearly had not taken this into consideration. Her advisors—her mother and husband—pointed to the number of million-dollar cases still in play without considering what happens to the bank offer every time one of those cases is eliminated. They kept talking about "safety shields," meaning that there are still a few million-dollar cases in play. But those shields prevent the bank offer from dropping to very low offers. Her best luck would have been to knock off as many $1 million cases as possible in the very beginning, for then each bank offer would have had a better chance at increasing with each round; then, her decisions for continuing would have been easier. The show designers had pulled off a good stunt. The subliminal message was that there is $10 million in play, while in reality the most that any contestant could take home would be $1 million—and that would have been almost as unlikely as finding a flying pig.

To appreciate the house effect, consider the following experiment.[8] Collect two groups, A and B, of random people. Give $50 to each student in group A along with the choice of doing nothing more or gambling by flipping a coin—heads wins $15, tails loses $15. At the same time, offer the students in B the choice of either taking $50 or flipping a coin with the condition that heads wins $65 and tails wins $35. A mathematician would not see a difference between the two conditions because the odds of winning are the same for both groups, whereas the psychologist would rightly predict that group A will have a higher percentage of students opting for the coin flip. Why? Students of group A will be considering the *house effect*; a loss is not a loss of personal money. But the students of group B are asked to gamble with partly their own money.

What constitutes a person's own money is not always so clear. For example, if you unexpectedly win a hundred dollars on a one-dollar lottery ticket, it may seem like an unaccounted windfall and therefore money that is given to you. Suddenly, you have money that is not a part of your regular earnings, so you feel freer to let it go. Your sense of monetary risk behavior is a consequence of prior gains and losses. And, fifty dollars or five million dollars, the psychological house effect is the same; risk decisions are made under the influence of prior wins or losses.[9]

Our little gambling experiment suggests two scenarios. Members of group *A* subliminally think, "I have already won $50. If I choose to gamble, there is a 50 percent chance of gaining $15 and a 50 percent chance of losing $15." Members of group *B* mentally account, "If I choose to gamble, I shall win $35 for sure, and in addition, will have a 50 percent chance of increasing my winnings from $35 to $65." For both groups, the earlier gain is larger than the potential loss, but the real difference is that group *A* is accounting the message in two stages—$50 is already won and there is a chance to make that $65— whereas group *B* has forgotten that previous stage of having just won $50; that is, the problem is posed as a pure gamble.

As we said, mathematically there is no difference between these two groups. However, the behaviors of each will depend on framing and encoding of the choices. In one experiment, 87 people were asked the following:

> Mr. A was given tickets to two lotteries. He won $50 in one lottery and $25 in the other. Mr. B was given a ticket to a single lottery. He won $75. Who was happier?

Though the winnings are the same for Mr. A and Mr. B, 64 percent felt that Mr. A was happier, 18 percent felt that Mr. B was happier, and 17 percent felt that there was no difference.

The same 87 people were asked:

> Mr. A received a letter from the IRS saying that he made a minor arithmetical mistake on his tax return and owed $100. He received a similar letter the same day from his state income tax authority saying he owed $50. There were no other repercussions from either mistake. Mr. B received a letter from the IRS saying that he made a minor arithmetical mistake on his tax return and owed $150. There were no other repercussions from his mistake. Who was more upset?[10]

Once again, the two events are financially equivalent, yet 75 percent answered that Mr. A was more upset, 16 percent that Mr. B was more upset, and 8 percent answered that there was no difference. This reflects a judgment that winning in two separate stages is more favorable than winning the same amount in a single event and that a two-stage losing is more upsetting than a single loss. Observe that each question could be reworded so that each pair of situations is not only financially equivalent but also hedonically so.

When it comes to human behavior and the management of risk, we find apparent contradictions. Behavior toward risk depends not only on how that risk is formulated but also on the risk taker's view of gains and losses. For example, a venture may be presented in terms of a risk-aversion or a risk-seeking experience. Suppose you were given the choice between a free gift of $1,000 and a chance of winning $2,500 by the flip of a coin. Which would you choose? The expected value of the second choice is $1,250. So if you chose to accept the gift of $1,000, then you are the kind of person that prefers to avoid risks. But here's the strange thing. In an experiment conducted by Daniel Kahneman (the Israeli behavioral economist who won the Nobel Prize in Economics in 2002) and Amos Tversky (the cognitive and mathematical psychologist who originated the study of irrational human economic choices and cognitive biases), two tests were conducted with a single group of individuals. In the first, everyone was given the choice between a free $1,000 and flipping a coin for $2,500 or nothing. In the second, everyone was given $1,000 with the proviso that they choose to either give away $1,000 or flip a coin for the chance of either having no loss or a loss of $2,500. It turned out that those who chose the free gift of $1,000 in the first test chose the more risky alternative in the second. Kahneman and Tversky discovered that when it comes to risk behavior there seems to be an asymmetry between risk aversion and risk seeking for mathematically equivalent choices.[11]

Take, for example, what economists call the Allais Paradox, named after the French economist and Nobel Laureate Maurice Allais, who conducted the following survey in the early 1950s.[12] Subjects were first asked to choose between two hypothetical possibilities.

A: A sure chance to win $1 million.

B: A 10 percent chance of winning $5 million, an 89 percent chance of winning $1 million, and a 1 percent chance of winning nothing.[13]

After choosing, the same subjects were then asked to choose between two additional possibilities.

C: An 11 percent chance of winning $1 million, and an 89 percent chance of winning nothing.

D: A 10 percent chance of winning $5 million, and a 90 percent chance of winning nothing.

Most people choose A over B and D over C. The surprise here is that the expected value of A is $1 million, whereas that of B is $1.39 million. The expected value of C is just $110,000 whereas that of D is $500,000. So what is the basis for such choices? It cannot be expected outcome, for there is an inconsistency. By expected value, B is favored over A and D over C. The surprise here is the irrationality of the reversal of risk behavior.

Much of gambling behavior rests on immediately prior gains or losses. For many, winning a hundred dollars at a slot machine is strong incentive for risking those same hundred dollars on a game with wildly skewed odds. There is a feeling that you are playing with the house's money and that it's free money to lose. Gambling choices are not made in historical isolation; they are made in the witnessing of former fortune outcomes.

For many gamblers, witnessing good fortune at the gambling table—whether it is theirs or someone else's—raises false hopes and encourages risky behavior, but neither such witnessing nor casino environments are fully to blame. The casino does not make the addict, nor does it make the risk seeker. It is still a mystery why some habitual social gamblers can manage their gambling pleasures while others lose all judgment of rational gambling behavior in thrill-seeking flirtation with jeopardy.

But it is that grip of the prospect of fanciful Fortune in the depths of enormous risk that compels some people to further adventure beyond the customary gambling games. Oblivious to the risks, they gamble under the mistaken belief that someone will give away something for nothing. Take the true story of John Worley, a fifty-seven-year-old ordained minister and psychotherapist who in June 2001 answered an e-mail request from one Captain Joshua Mbote of

Nigeria.[14] In the early years of this new century almost everyone with a computer connected to the World Wide Web got the same message from Captain Mbote.

> My name is Mr. Joshua Mbote. . . . I was the president of Ghana petroleum cooperation. . . . I have made up my mind to evacuate and invest 35 million dollars abroad . . . it will be highly appreciated if you can help me to achieve my aims. . . . Please be rest assured that this transaction is 100% risk free and must be treated with utmost confidentiality it deserved . . . waiting for your urgent response.
>
> Best regard
> Joshua Mbote

The sensible among us would have no doubt that such a message is fraudulent. So how could John Worley—a man who preached against Satan at his granddaughter's graduation, "He's going to be trying to destroy you every inch of the way," a man who had spent so much of his adult life preaching self-reflection—have been fooled? Surely he knew that his e-mail had come from satanic criminals who could destroy him. Yet there was some combination of wishful thinking, avarice, and faith in God's will that drove him to answer that first of many e-mail messages.

Helping Mbote would not be quick and simple. There would be phone calls and e-mails saying that if he were to cooperate, he would receive more than six million dollars by being a middleman for the transfer of funds. But he—a man who had already declared bankruptcy and had little personal savings for retirement—would have to shell out an up-front cost for transportation and incidental charges. Was this not gambling? For the next five years John Worley would give Mbote over $40,000 of his own money in advance fees and cash $648,500 of Mbote's phony checks. The stakes were high; besides the highly risky financial gamble, deliberate bank fraud is a felony. Though he knew that he was dealing with fraud—bribing Nigerian bank officials and telex operators—his vested interest in the scheme continued to deepen. Easy money was in greed's sight.

The Nigerians turned Worley's skepticism into such a suspension of disbelief that Worley began to worry that *they* might not trust *him*. They transformed him into the perfect mark.[15] In the end, Worley was sentenced to two years in federal prison plus full restitution of over $600,000. In one of his e-mails to the Nigerians, he wrote, "I have been taken advantage of by you evil bastards. . . . I am ashamed, and shamed, and an embarrassment to my family, who are so precious and Godly people. . . . Thought[s] of suicide are filling my mind. . . . I hate living right now, and I want to die. My whole life is falling apart, my family, my ministry, my reputation and all that I have worked for all my life. Dear God, help me. I am so frightened."[16]

Even after his prison sentence, Worley received another message from Mercy Nduka, one of the Nigerians: "I am quite sympathetic about all your predicaments," she wrote, "but the truth is that we are at the final step and I am not willing to let go, especially with all of these amounts of money that you say that you have to pay back." She needed just one more thing from Worley and the millions would be theirs: another three thousand dollars.

"You have to trust somebody at times like this," she wrote. "I am [a] waiting your response."[17]

Worley may have been the Nigerians' best mark. His story may be extreme. But for every one extreme, every one fattened for the big kill, there are hundreds of bony marks fallen to cons.

♣ CHAPTER 13 ♣

The Theories

What Makes a Gambler?

The next best thing to gambling and
winning is gambling and losing.
—*Nick "The Greek" Dandalos*

Late nineteenth- and early twentieth-century theories of psychopa-
thology did not distinguish between so-called neurotic gamblers
and social or entertainment gamblers. Starting with the Freudian
school, such psychoanalytical theories were based on the assumption
that almost all of human behavior is a result of subconscious mental
activity. Added was the *collective unconscious* with its instinctive memories and inherited diary of neurotic tensions stored in the id, transmitted generation to generation from the time of our earliest ancestors,
what the nineteenth-century gambling historian Andrew Steinmetz
called *hereditary transmission*. Opposing the id is the superego, a place
in the subconscious holding parental reproaches, moral censorship,
and cultural conventions. The ego, answering to man's nature, tries
to mediate the conflict, but as these powerful internal conflicts brew,
the individual is plagued with feelings of guilt and anxiety, which
Freud considered the surfacing of the id over the superego as a tension between the conscience and the actual performances of the ego.
It was believed that that tension is experienced as a sense of guilt.[1]

Freud's idea of these internal conflicts came from ancient myths purporting intense incestuous acts, such as the story of the Theban king who unknowingly killed his father and married his mother; those myths, according to Freud, build in the collective unconsciousness of human phylogeny.

Applied to gamblers, these theories say that the gambler's true motivation may not be his conscious will to win but an unconscious desire caused by some internal conflict, possibly even an unconscious desire to lose. To us, this may seem radical, outrageous, and perhaps the twaddle of an idea made into theory without evidence. Nonetheless, the theory is intriguing, for it connects the gambler's desire to control that which cannot be controlled with the unconscious wish for punishment. In the foreword to his book, *The Psychology of Gambling*, the distinguished twentieth-century Freudian analyst and clinician Edmund Bergler wrote that the purpose of his book was to prove that the gambler has an unconscious wish to lose, and that that is why the gambler always loses in the long run. Gambling is not played for money but as a proxy for the pleasure of released lustful drives and aggression rewarded by guilt-loaded punishment in the form of lost money.[2] The gambler lusts for punishment. In German, the word is *angstlust*. He knows that his loss is guaranteed by unfavorable odds, yet he continues his masochistic thrills dictated by the pleasure principle, which in his case is pleasure from displeasure.

Freud may have blamed Dostoyevsky's gambling on masturbation guilt and self-punishment to ease that guilt, but other psychoanalysts had other ideas. Indeed, Bergler, for instance, claimed that the gambler has an unconscious desire to lose as a self-punishing psychic masochism reaction to the guilt-driven Oedipal conflict, an unconscious hostility toward parental figures, which may include other authority figures such as teachers. The Grandmother in Dostoyevsky's *The Gambler* watches a very young man at the roulette table winning heavily. She sees his eyes flashing and his hands shaking, and notes that he continuously staked without any sort of calculation, winning and winning, raking and raking in all his gains.[3] The young man continues to stake and rake his winnings. And then we are told that the Grandmother suddenly turns to Alexis Ivanovitch, the narrator, and breathlessly exclaims, "Go and tell him—go and tell him to stop, and to take his money with him, and go home." To the consternation of everyone, especially the croupier, she shouts, "Presently he will be losing—yes, losing everything that he has now won . . . the young man is done for! I suppose he WISHES to be ruined. . . . What a fool the fellow is!"[4] The Grandmother seems to know all about psychic masochism, pleasure from displeasure—"he WISHES to be ruined," she says. And as we read the pages, we wonder if he ever leaves the table with anything. We are not told, for the story continues with the Grandmother's escalating excitement of the game.

If belief in luck is less the wish to win than the opportunity to control what cannot be controlled, then the gambler consciously thinks he or she has control of the outcome, that he or she will win simply because there is a will to win. Yet, if the psychic masochism school is right, he or she unconsciously wishes punishment to ease his or her guilt. The pathological gambler consistently loses more than he or she wins; yet often those losses do not diminish his or her belief in control.

Listen to Alexis Ivanovitch's belief in luck through a conversation between Polina Alexandrovna, the young lady with hopes of an inheritance from the Grandmother. It is fictional, not a clinical account; yet Alexis's narrative does illustrate the compulsive everygambler's strong illusion of control. He had just gone through a moderate loss in betting for Polina.

"I always felt certain," he tells Polina in a gloomy tone, "that I should win. Indeed, . . . [I] ask myself—Why have my absurd, senseless losses of today raised a doubt in my mind? Yet I am still positive that, so soon as ever I begin to play for myself, I shall infallibly win."

"And why are you so certain?"

"To tell the truth, I do not know: I only know that I must win—that it is the one resource I have left. Yes, why do I feel so assured on this point?"

"Perhaps because one cannot help winning if one is fanatically certain of doing so."[5]

Freud felt that certain obscure laws of human behaviors are independent of culture, time, or race, and universal qualities like vision
or appetite. The demonic impulses that connect Hamlet to Oedipus
may very well be a poet's way of personifying the ego's subtle struggles with the real world. And so, Freud thought that pathological
gambling parallels a subconscious desire to self-castigate.[6]

This may seem a stretch to the behaviorists out there, who believe
Freud's theory of gambling has little merit in the light of end-of-
century theories of cognition, but at the time it was something new
and, like all new theories, spawned introspection, discussion, and an
explosion of surrogate ideas.

Many early twentieth-century psychoanalysts followed Freud's
analysis of gambling, some stretching the Oedipal features to autoerotic desires. Bergler believed that when gamblers place their
stakes in the attendance of luck, they are quixotically commanding
a win, all in the illusion of omnipotence. Bergler would say that an
addictive behavior, especially irrational gambling, is a test to control
Fortune.

> The conscience is like a sadistic jailer who drives his prisoner to
> such desperation that he beats his head against the wall. At first
> the jailer gloats, but at the first suspicion that his victim is deriving
> pleasure from this self-torture, the jailer will force him to stop.[7]

Another prominent early twentieth-century psychoanalyst, Hans
Von Hattingberg, connected gambling obsessions to the toddler
years of toilet training, when unrestricted bowel eliminations (which
he believed to be an autoerotic pleasure) suddenly become checked
under toilet protocols for the first time. This discouraged pleasure
emerges in adulthood as the obsessive desire to gamble—the unconscious views this as the unrestricted flow of cash.

Of course, there were many more early psychoanalytic views on the
nature of gambling, almost all based on subliminal grounds, many
interesting from an academic point of view, and many took a clinical approach. But none supported resounding scientific evidence.
Neither is there modern scientific support for theories of inherited
instincts, masturbation drive, Oedipal impulses, or masochism.

Testing such theories was (and still is) difficult and expensive. Many early theories were tested on either very small sample sizes or biased samples. So what is the current thinking? Unfortunately, it is still all theoretical and inconclusive. All we can do is categorize according to camp. In the mid-1950s the leading belief was that the causes of excessive gambling could be found within the individual psyche with little credit to environmental factors. Out of Bergler's 200 cases, 80 were acute, 60 continued treatment, and 45 were completely cured (meaning that they stopped gambling, recognized their inner conflicts, and ended patterns of self-destructive behavior). Yes, Bergler cured roughly two-thirds of those sixty problem gambling patients who continued treatment, using purely psychoanalytic techniques, but some behaviorists argued that the high number of cures could have been circumstantial, that perhaps the mere clinical attention resolved their problem. Other psychoanalysts felt that many patients with multiple neuroses were convinced that gambling was a problem only after lengthy analysis. Moreover, because Bergler never mentioned any follow-ups, no one ever knew if those cures were permanent or temporary.

Though there were other reports of clinical successes with patients, behavioral psychologists developed suspicions that environmental factors could play more than just a derivative role, even perhaps a fundamental, key role in determining motivation for gambling and other addictions. Even Edmund Bergler conceded that he saw one thing common to all the addict gamblers he treated in his thirty years of practice: that gambling is just one of the many unconsciously self-created, and self-perpetuated, self-damaging tragedies in their lives.[8] Yet, at the same time, he believed that everyone in Western cultures is a potential gambler, possibly harmless or possibly dangerous.[9] This view does not represent psychoanalytic orthodoxy, as it suggests some element of personality influence. He also claimed that certain personalities have dormant tendencies that can awaken at any time to activate the gambler within.

Shocking, eh? Yes! Was Dostoyevsky any different than Tolstoy? They were both gamblers. In his youth, Tolstoy had to sell parts of his estate to pay off huge gambling debts. With a 1,000-ruble gambling

The terms *pathological gambler* and *problem gambler* are distinguished in this book by the nomenclature used in the 1994 edition of the DSM (*Diagnostic and Statistical Manual of Mental Disorders*) of the American Psychiatric Association. *Pathological gambling* refers to "a mental disorder characterized by a continuous or periodic loss of control over gambling, a preoccupation with gambling and with obtaining money with which to gamble, irrational thinking, and a continuation of the behavior, despite adverse consequences." *Problem gambling* refers to "gambling behavior that results in any harmful effects to the gambler, his or her family, significant others, friends, coworkers, etc." The distinction rests with the harm to others factor. In popular literature, the term *compulsive gambling* is synonymous with *pathological gambling*, but here we avoid the term because, according to the psychiatric lexicons, the word *compulsive* refers to "repetitive behaviors or mental acts, the goal of which is to prevent or reduce anxiety or stress, not to provide pleasure or gratification." Compulsive behavior is a challenge to relieve oneself of discomfort. This includes performing, idiosyncratically, acts governed by self-imposed irrational rules. Typical pathological gamblers, in contrast, gamble for pleasure, though one could argue—as Freud has in his analysis of Dostoyevsky's gambling—that some forms of self-imposed pleasure are auto-gratification and hence compulsive.

debt over billiards, Tolstoy had to relinquish his unfinished manuscript of *The Cossacks*. As Jerome Charyn says in his introduction to *The Cincinnati Kid*, "We are all players of one sort or another. . . . Defeated like the Cincinnati Kid, we wait, wait, wait until the dream comes back to us, a bit at a time."[10]

Though the Grandmother's appearance in Dostoyevsky's *The Gambler* is mostly a diversion, the final outcome of her gambling escapades portends the ruined romance between Alexis Ivanovitch and the General's stepdaughter Polina. We watch this character of a woman who at first appears to be arrogant and cantankerous and yet, against class propriety, insists on entering the casino with her old servant Potapitch and her maid. "Rubbish! Because she is a servant, is that a reason for turning her out? Why, she is only a human being like the rest of us."[11] This is how she feels before leaving for the casino. But once the gambling takes over, she abandons kindness to her servants and harshly dismisses them. "Why have YOU attached yourselves to the party?" she tells her servants on her way to the casino for the second time. "We are not going to take you with us every time. Go home at once."[12] Joseph Frank, the preeminent Dostoyevsky biographer and commentator, noted that as the passion for gambling gains strength, it begins to dominate all other human feelings. In the passion, all relations between humans diminish and in some cases cease to exist.[13]

Many literary commentators have identified Alexis Ivanovitch with Dostoyevsky as a rendering of the start of Dostoyevsky's gambling craze.[14] It is true that Dostoyevsky's financial position was in serious trouble; the Russian government censored *Vremya* (Time), the magazine he and his brother Mikhail owned and published (a publication that was a leading literary journal in Russia) because it allegedly supported the Polish revolt of 1863. After receiving a generous inheritance from a wealthy aunt, he launched a new journal, *Epokha* (Epoch), with his own money, but after Mikhail died, the debts of *Vremya* to creditors were high and Dostoyevsky's financial situation collapsed completely. He had taken on the debts in the form of promissory notes. Threatened with debtors' prison (which the law allowed in those days) and reduced to despair and desperation in the summer of 1865, he turned to Feodor Timofeevich Stellovsky, a cunning, unscrupulous publisher, who advanced him a sum of three thousand rubles in exchange for the right to publish Dostoyevsky's complete works. Risky and excessive though it was, the agreement included the right to publish all future works without any compensation for a period of nine years, if Dostoyevsky did not complete his new novel *The Gambler* by November 1, 1866.[15] He agreed, but the money went straight to his debtors. Just after this agreement, he got an advance from another journal for travel articles. So, with some money in his pocket, he left Moscow for Europe.[16]

Under the circumstances we are not surprised to find Dostoyevsky visiting the gaming rooms of Homburg, Baden-Baden, and Wiesbaden in desperate attempts to regain his losses. His wife, Anna, recounts his time in Homburg: "After a few days I began receiving letters from Homburg in which my husband told me about his losses and asked me to send him money. I did so, but as it turned out he lost that money as well and asked me to send more, which I of course did."[17] He told her that he could feel his luck when playing roulette, but that his luck could not be sustained because his own impatience would cause him to change his bets and systems (which he thought foolproof) and that he had too little money to sustain unfavorable turns. Those were his reasons for losing. He was so convincing that even Anna was persuaded of his gambling theories and willingly permitted him to go to Baden-Baden for a week. That week turned into two, then five. In her diary, Anna concluded that some kind of nightmare endlessly possessed her husband.

Each week Dostoyevsky would turn to the tables and lose everything. Each week he would find a way to get more money to go on gambling, chasing his losses. And it would continue till he had lost everything he had. Like many pathological gamblers, he was convinced of his theories. Some theories may have been as foolproof as he had claimed, but they all depended on having a substantial amount of money—which he did not have—to bring him through the adverse turns of the wheel. He would return in the mornings from the roulette tables with nothing in his pockets and, crushed, head to the pawnshops, hardly able to stand on his feet, to pawn the few valuables left in his meager apartment.

Within behavioral psychology there is personality theory, which presumes that an individual behaves in accordance with his or her personality type—a compassionate type would be inclined to volunteer for a homeless shelter unconditionally, whereas a self-centered type would need a reward. In the behavioral camp the question remains whether or not central personality is genetically determined or dependent on life's experiences. For example, some psychoanalysts pointed to early childhood problems with parental approval and the unconscious perception of denied love; the gambler turns to luck for love and acceptance.[18] Personality theory attributes gambling urges to environmental influence, not to emotional trauma in infancy, so consciously learned habitual responses to external stimuli may appear at any stage of life. The clinical advantage is that it is possible to unlearn what is learned at any stage of life. Some psychologists within this group believe that repetitive behavior is reinforced by rewards. (For gambling the rewards are monetary or emotional.) The behavioral psychologist B. F. Skinner believed that problem gamblers are victims of variable reward schedules (random wins and losses) and therefore persist at the game for longer periods after the reward is withheld.[19] Professional gamblers, therefore, are able to manipulate novice gamblers by varying the rewards, purposely permitting wins in the beginning and cleverly aiming for losses in the end.

This is not meant to be a censure of professional gamblers. Most are honest and use skills in accordance with the broad scope of the game. The honest professional knows who to play and generally plays with someone whose ranking is slightly below or above his or hers.

Behavior is most often reinforced by our immediate physical surroundings and their consequences, and such reinforcement is regularly repeated.[20] Most often we know when to eat and when to sleep, what we like for breakfast, which road takes us to work and which road home. This is a good thing; it stabilizes our feelings for the everyday world around us and gives us enough comfort to balance our emotions and coping skills. The professional gambler leads his or her victims on by adjusting reinforcement and control; letting the victim win at first, then, slowly or rapidly, depending on how long the professional intends to play, increasing control to slow down the winnings.[21] Take, for example, the slot machine. In that case the professional is the house and the victim is the player. The player knows that if three copies of the same picture appear, then jackpot. So when any two of the same picture appear—the third different—the

reinforcement is *Ah, I ALMOST got the jackpot!* That *almost*—an almost
that cost the casino absolutely nothing—increases the likelihood
that the gambler will continue to play.

Conditioning is also a powerful player.[22] We know from Pavlov's
experiments that if you present a hungry man a dish of food often
enough, the man will salivate at the sight of an empty dish. It's as if
the dish has acquired a property, the power to stimulate a behavior.
Replace the food with money and we have paired a conditioned reinforcement of the hungry man's salivation with the acquisitive man's
drool.

Gambling payoffs are reinforcements of this unpredictable kind.
For gamblers, the inclination to bet depends on an unpredictable
schedule of reinforcement. Casino owners know this and adjust the
odds on a varying schedule of reinforcement to generate gambling
behavior in their favor. They simply select a mean ratio between two
extremes of reinforcement and control. For example, they sponsor
slot machine tournaments with prizes to lure in a crowd of willing
novices along with a few fervent enthusiasts. For them, there is nothing unsafe or unpredictable—their long-term profit is calculable.

The behavior of the gambler is under very complex control
depending on his or her history of reinforcement. Moreover, vulnerable gamblers are often guided by cultural exposure—neighborhood
crime levels, mental health, and family customs, as well as proximity
to and convenience of gaming. Reinforcement and conditioning—
the significant elements of complex control—motivate the gambler's
decisions. At the psycho/economic level, the consumer hands over
money for goods by conditioning and reinforcement just as well as
by needs and bargains, and, so too, the gambler places his or her
stakes on bargains of significant value—the greater the bargain, the
greater the probability that he or she will make a bet, and hence the
greater the reinforcement effect through winning or losing. The betting interest might depend on the stake size as well as the gambler's
reinforcement profile.

Casino owners know that the longer the game continues, the
larger the bets. Large rooms, noisy machines, sounds of spilling
coins, hubbubs of crowds, and multimedia presentations, along with

the smells of food, and perfume, provide not only Pavlovian vibes to encourage gamblers to stay with their games for as long as possible but also distractions, such as flashing lights and free drinks, feeding cognitive overload to reduce rational cognizance and to promote guessing and careless gambling.[23]

Cardsharps and professional gamblers know what is called the *early win hypothesis*: that their opponents' early fortune fosters belief in continued good fortune. Gamblers who win early believe that any subsequent losses are temporary and that those losses will be conquered with persistent betting. The logic here is that quitting means a sure loss, whereas persistence in chasing the lost money may lead to winning, or at least the possibility of recouping and breaking even. So, like foolhardy politicians who start reckless wars they can't possibly win and who—too late and too far gone—are forced to spinning *defeat* into *leaving with honor*, they chase their losses. Increasing the size of their wagers, they inevitably continue to lose in the long run to become entrapped beyond the rational quitting point when they could still keep their losses to affordable levels.

Behind the scene is the schedule by which the gambler will win. But for every winner there must be at least one loser. Typically, there are many more losers than winners. If the jackpot of a slot machine were one dollar, the casino—assuming exceptional generosity—could manufacture machines that would pay out the jackpot on, say, one out of eight tries. If it costs a quarter to play, then, on average, after every four tries you might lose a dollar, but then again, with luck, you might double your money. Thinking along these lines, why not have a state-sponsored casino where everyone wins, so everyone can be happy—the money simply changes hands from one gambler to the next.

Yes, casinos could do this and so could state lotteries. But studies have shown that participation tapers off with the pot. We—the generic *we*, of course—are far less interested in winning a million dollars than a hundred million. The states could substantially increase the chances of winning jackpots by intentionally splitting the winnings a hundred times by selecting a hundred different jackpot-winning numbers.[24] However, like casinos, the states are in the gambling business to make money to offset their revenue deficits, and therefore increase the gambling pool.

When we add Pavlovian conditioning, the reflex to specific stimuli, we find the poor gambler controlled by rewards associated with gambling: a croupier's call, the sound of clicking chips, the spinning of the wheel, the shuffling of cards. The arousal, excitement, tension, and anxiety all contribute to the onset of addiction. It happens to Alexis Ivanovitch in *The Gambler*—"Even as I approach the gambling hall, as soon as I hear, two rooms away, the jingle of money poured out on the table, I almost go into convulsions."[25]

However, there are problems with behavioral theory as well. Like psychoanalytic theory, it is difficult to get a real scientific measure of success. And some psychologists are not convinced that the general pathological gambler is psychologically any different from those with other addictions.[26] Current neurobiological research indicates that addictive behavior is not likely to be isolated in any particular addiction but is rather entrenched in complex biological, psychological, and social underpinnings. Excessive behaviors such as alcohol abuse and pathological gambling may not be independent disorders.[27] Indeed, the Gamblers Anonymous twelve-step program is almost identical to the Alcoholics Anonymous twelve-step program.[28]

The good news is that large sampling statistics now tell us that neurotic gamblers are relatively few in proportion to the vast number of all gamblers. According to the National Research Council 1998 report from the Committee on the Social and Economic Impact of Pathological Gambling, 1.5 percent of the U.S. adult population were at some time in their lives problem gamblers, and 5.4 percent were either problem gamblers or pathological gamblers at some time in their lives. These may seem surprisingly low percentages after considering the percentage of Americans who gamble. According to a Gallup survey conducted in 2004, two-thirds of all Americans had gambled at some time during 2003, though an overwhelming percentage (74 percent) of them gambled by purchasing lottery tickets. There are more than 890 casinos in America with combined gross revenues topping $52 billion. According to a 2006 Pew Research survey, nearly 29 percent of all Americans visited a casino at least once that year, an increase of about 20 percent from the previous decade. Unfortunately, most surveys are flawed by the problem of artificial answers. Though so many Americans do gamble, few will ever say that their gambling is the source of problems with family, friends, and work. A survey conducted by Gallup USA in 2003 asked, "Has gambling ever been a source of problems with your family?" Just 6 percent said yes. To get accuracy, the best source is Gamblers Anonymous itself, which points out that among its members,

78% suffer from insomnia
26% had been divorced or separated because of gambling
34% lost or quit their jobs
44% had stolen from work to pay gambling debts
18% had gambling related arrests
66% had contemplated suicide
11% had attempted suicide.[29]

We don't know the percentage of successful suicides due to gambling, as those who were successful were not there to participate in the survey. Hearsay about gambling-promoted suicides is plentiful. I recently heard this curious heartbreaking story from the manager of a Camelot betting shop in West London. One day in 1995, a year after the British National Lottery was established, an unkempt man walked into a betting shop in Ealing with a suitcase, placed six crisp £50 notes on the counter, and asked to buy 300 Lotto tickets. The jackpot was up to £7,500,000 that week. The man started off with his entire life savings of £23,800 in his suitcase. He had located 200 betting shops in the UK that had never before sold a winning Lotto ticket and randomly picked 80 to distribute his ticket purchases. The shop at Ealing was one of the last on his trip. Aside from his disheveled look, there was nothing peculiar about the man, but when asked how he would feel if he did not win, the response was, "I will kill myself." At that he left the shop. Nobody won that week.

Some behavioral psychologists believe that gambling addictions are reinforced by a driving need for stimulation as a self-medicating means of coping with boredom. And some believe that betting, rather than winning, gives the emotional kick. Anyone who has closely observed gamblers at the start of a race will mark the gambler's boosted exhilaration just before a race begins. A theory of sensation-seeking is supported by such an observation, when bets are placed very close to the start of a race—the closer to race time, the higher the exhilaration. At Monte Carlo I observed one cigar-smoking gentleman play seven hands of blackjack at once, consistently winning almost every one. His table was roped off so I could watch from a distance only. Why didn't he go to the utterly private rooms, where Saudi princes play without a limit and without being seen? Because he *wanted* to be seen.

For many, it's not the money, entirely. It's the action, the euphoric state, the rush. I observed several women at Monte Carlo ritualistically placing their bets at the roulette wheel only at the very last moment, just before the croupier would call out *pariez non plus*. One giddy middle-aged woman sat at the blackjack table for the pure pleasure of gambling. Her wrists were covered with diamond bracelets. From her neck hung three strings of pearls. She began the evening winning at blackjack, collecting stacks of chips totaling almost €8,000. But at the stroke of midnight, shocked at the time, she announced, as if a pumpkin carriage was waiting for her outside, that she had to go. At that moment she played her entire pile of chips and, within two minutes, lost. She tipped the croupier and left with the same carefree gaiety she had exhibited all evening. She had her pleasure. A young man sitting next to her hit blackjack several times. When he did, his head would roll back, and his eyes would lift above his lids as if in a swoon. It was more than blissful.

Pathological gamblers have described the feeling to clinicians as similar to a cocaine high, complete with cravings and the urge to take bigger risks for greater excitement. Still, for others it is not the excitement that drives them but the escape, a numbing experience designed to reduce whatever mental discomforts they may have; amnesia, trances, and altered states of being have been reported.[30] And still others have been clinically observed to cognitively distort and deny reality in favor of superstition to enhance their control of luck and chance. For example, they misrepresent odds and brand particular numbers as lucky. One woman at the casino in Nice would consistently squint and give her head a slight twist while looking up at the electronic board that listed the previous ten outcomes just before placing her bets.

Some pressures are mixtures of cognitive and social factors. They include acquired behaviors through social influences of peers or role models. Some gambling may be explained as simply a lifestyle formed from self-medication that combats anxiety, insecurity, and low self-esteem, misrepresents the self-image, and excuses, rationalizes, and justifies extreme behavior, all the while giving the addict an indomitable elevated feeling.[31]

Another theory claims that gambling satisfies a physical or emotional need caused by some (innate or nurtured) internal call to escape the routines of reality, a driving call for recreation and amusement. In this theory, excessive drinking, drug abuse, and gambling are acquired bad habits developed to cope with the routines of life; any therapy that does not completely quench those coping responses will lead to relapse.

Some gambling conditions cross over several theories and some are too baffling to peg. When I first went to the Monte Carlo casino I was underage and was permitted only as far as the grand entrance lobby, which, at that time, had a foyer off to the side filled with slot machines. It was early morning when the cafés outside were busy with some people having their coffees and croissants and others their morning cognacs. Inside the palatial hall were the players, not the high rollers who would have left in their Rolls Royces and Maseratis long before daylight, but the weary players who had been up all night playing roulette or blackjack, or pulling levers on those one-armed mechanical slot machines. One woman in particular was of interest to me. From a distance I watched her pull out a franc from her small bag, put it in a slot, pull the lever, and, not waiting for its reels to slow down, frantically move on to play the next machine. She continued in that way down a long aisle of machines.

For years, that woman's face lingered in my memory and every time slot machine rooms at casinos came to mind, whether from reading novels or doing gambling research, I thought of her. No words were exchanged. I could not understand why she did what she did, and attributed her behavior to some strange gambling neurosis beyond my grasp. And then one recent day in Nice I saw another woman do the same thing. It was euros this time. From a small sack of euros she fed a row of machines, one by one, without waiting for reels to slow. It was as if she knew in advance that she would lose. This time I questioned the lady. Taking a closer look, I could see the thick powder on her face, the blue eyeliner and thick mascara, the bloodshot eyes, the despair. She must have been in her late sixties.

"Tell me," I said, "why not play just one machine?"

At first she didn't reply, but as I excused myself for interfering and turned to leave, she called to me from a distance and said, "You don't know which machine is due, so your best shot is trying them all."

That thought had never occurred to me, but when I looked around I saw other men and women doing the same thing, moving from machine to machine with sacks of euros. I spoke with others.

"I get like this whenever I come into this place," one woman, reeking of alcohol said. "It's the lights, the drinks, the noise. It makes me giddy."

"Is it the entertainment?" I asked.

"No—yes, well, maybe, I don't plan to throw money at the slot machines. I play a little roulette and whatever coins I still have at the end of the night go to the slots. I can't pass a slot machine without a try. I wish I could, but can't."

The striking thing about all those people was that every one of them had a desperate, unhappy look, a forlorn look, as if they had just lost someone or something serious in their lives.

And finally there is a cognitive behavioral theory, which suggests that some addictions are the consequences of an irrational mind. These include the logical fallacies about winning numbers, choices that result from biased conclusions from erroneous grasps of probability. Like the Grandmother in *The Gambler*, lottery and roulette players will continue to play the same numbers over and over. Many gamblers are convinced that luck is a personal attribute that can be predicted and manipulated through magical thinking or superstitious behaviors. Thus they touch their noses or wear a particular garment or talk to the dice or machine—lucky ties, jewelry, furry dice, and so forth.[32] They rely on omens such as license plate numbers, radio reports, or birthdates. Sports gamblers favor specific athletes, teams, or horses based on names, hopes, or other groundless criteria rather than on some hard evidence of who or what is most likely to win. Their winnings have no justification, their losses defended as flukes. Humans have selective memory. But the same mechanism that is intended as a blessing—a natural defense against the worst tragedies—works against the gambler who recalls past winnings and forgets all losses—a denial that accelerates gambling moods. (According to a 2006 Pew Research survey, 56 percent of gamblers say they are ahead for the year.) On top of that, the irrational gambler has confident illusions of control over the outcome—in the way the dice rolls off the palm of the hand, in the choice of number, in the feeling of having luck.

Throw a lucky man into the sea and he will come out with a fish in his mouth. So goes the Arab proverb. But we should look at the notion of luck more carefully. It is true that some games have varying degrees of luck and skill, but we should keep in mind that chance dominates almost all the obvious games, excluding pure skill ones—such as chess—at the far end of the chance/skill spectrum. Though poker is well-known to be a game mostly of skill, honest players need some luck to get a good hand to work with.[33]

There is no one-size-fits-all theory. The slot machine gambler may be motivated by the principles of behavior whereas the dog racer may be encouraged by something far more cognitive.[34] Structural forces begging for a wager bombard the addict, or potential addict. First, gambling evokes financial rewards and emotional excitement. Both are cognitive causes, but some psychologists sense the undercurrents of both Skinnerian and Pavlovian views. Next, the gambler is partial to the highly publicized wild winnings of others with the thought that they can be duplicated. Bergler once wrote that every gambler recalls stories that allegedly prove it is possible to get rich by gambling. What is interesting is that almost all such stories are frequently known through hearsay.[35]

An AP news item on January 24, 2008, relayed the story of Sammy Zabib, a forty-two-year-old New York limousine fleet manager who sat down at a Brazil Slingo slot machine and played at $4 a spin for about an hour. Zabib won the jackpot of $779,000, and for the next few days, while newspapers around the country from the *New York Daily News* to the *Seattle Times* (along with network radio news and NPR's *Morning Edition*) carried the story, Atlantic City casinos were abuzz with Brazil Slingoers.[36] If any did get to the roulette wheels, some would feel the pull of the Monte Carlo fallacy, that infamous gambler's mistake that whispers into the novice's ear at roulette wheels everywhere, *after a long run of red, bet on black*. Publicized large-winning success stories reinforce the gambler's persistence in the face of hefty losses, which inevitably compound gambling dependence. Such success stories reinforce belief that the spinner controls where the little steel ball falls and so bets are placed with an aim to outwit the spinner.[37] Out of those publicized large-winning stories come teams that believe that they can beat the system because they have infallible systems and special skills to give them advantages over other gamblers. Somehow, they believe that there is a skill in picking the outcome. They excuse and forget their losses as caused by things beyond their control, and count their winnings to prove their systems work.[38]

What makes a person a gambler? Is it the lust for risk, the desire to control destiny, something in the collective unconscious, inner guilt, yearning for punishment, or any other of a mass of answers? The most recent research, using PET scans, suggests that pathological gamblers, alcoholics, and drug addicts have similar patterns of neural activity when exposed to their individual addictions. We know that just as the sight of drinks easily seduces alcoholics, lottery drawings, casinos, and Internet gambling sites strongly influence pathological gamblers. PET scans of pathological gamblers show increased levels of dopamine during play and even more substantial increases during high-risk, high-stakes playing. When neuroscientists looked at a particular group of neurons in the ventral tegmental area (VTA) at the very center of the brain, they found that circuitry associated with pleasure increases during activity. It is the pinpoint location of pleasure and addiction processing, a center of control for both reward experiences as well as the emotions connected to those experiences. This is where information is received from the cortex about the satisfaction of fundamental human needs such as food cravings and other assorted sensual comforts. By increased activity, the VTA releases dopamine, a chemical messenger, which gets transmitted to the nucleus accumbens, a collection of neurons within the forebrain known to play a major role in reward, fear, pleasure, laughter, and addiction. With increased and prolonged activation those neural pathways of dopamine transmission (reward pathways) are strengthened to favor pleasure, reward, and punishment, as well as other distinguishing behaviors of the individual, thereby reinforcing those behaviors.[39]

However, dopamine production also occurs during many types of high-anxiety brain activity such as repetitive complex tasks that become increasingly difficult. It is also triggered by the pleasures of eating, sex, alcohol, nicotine, and other stimulants such as drugs, sugar, and chocolate. In other words, transmission does not distinguish addiction type. In fact, dopamine transmission does not differentiate the activities of extensive gambling, obsessive drinking, and so forth. So we must distinguish between the measurement of increased dopamine and the cause. There is much to learn using these new sophisticated technologies. And yet many studies tend to revive some variations on Bergler's earlier theories of self-punishment. Joseph Frascella, the director of the Division of Clinical Neuroscience at National Institutes of Health, claims that addictions are not only persistent behaviors facing negative consequences but also the unconscious desire to continue something you know does harm.[40]

The latest research observations are not so simple as to be limited to dopamine activity in the VTA. Most addictions are tightly connected to unrelated multifarious psychiatric complexities involving long-term tendencies, habits, and environmental conditions, as well as continuously active libido energy.[41]

There is still so much that is not known; yet neuroscientists are beginning to focus on the effects of neurotransmitter imbalance in designing new drugs that may inhibit the transmission of dopamine during cycles of detrimental addictive behaviors. The trick is to separate destructive behaviors from biologically necessary ones without purging the prudent delightful pleasures of life.

Reasons for gambling obsessions are complex and connected to the multifarious complexities of normal human behavior. If all addictions were the same, then it should seem that treatment would be easy, for we should expect that the solution for one would be the solution for others. However, it turns out that stopping one intensifies another. We see ex-smokers compensate their withdrawals by overeating and ex-gamblers turn into alcoholics. A trip to the store for a quart of milk may turn into a stop at the liquor store. A drink or two lowers inhibition for an innocent gambling rendezvous, and pretty soon the ex finds him- or herself compulsively gambling at a casino or online.[42]

♣ CHAPTER 14 ♣

Hot Hands

Expecting Long Runs of the Same Outcome

Should I go to heaven, give me no haloed angels riding
snow-white clouds, no, not even the sultry houris of
the Moslems. Give me rather a vaulting red-walled
casino with bright lights, bring on horned devils as
dealers. Let there be a Pit Boss in the Sky who will
give me unlimited credit. And if there is a merciful
God in our Universe he will decree that the Player
have for all eternity, an Edge against the House.
—*Mario Puzo, Inside Las Vegas*

Remember how the Grandmother in *The Gambler* won eight thousand rubles through her luck of the ball falling on zero three times in succession? She stopped playing when she was ahead, but, with all the fuss over her, she was compelled to return to the casino the next day and hastily take her place next to the croupier. In chapter 12 we hinted that moments of glory pulled her back to the table for greed to take over. Eight thousand rubles was a tease, and so easy to win! So why not double it? We are told that the casino owners and croupiers expected her to return, for the croupiers of Roulettenberg saw the Grandmother as permissible prey.

It was now the croupier's turn to win back some losses. The old woman still thought zero to be her lucky number, and so for each of the first twelve rounds staked a ten gulden coin (the value of pure gold the weight of approximately six sugar packets) on zero. Of course, zero never turned up. Giving up on zero, she had Alexis Ivanovitch stake four thousand gulden (almost 5 pounds of pure gold, the maximum permitted for any one time) on red, and that's when zero turned up. At that point she had Alexis Ivanovitch stake another four thousand on red. She lost again. There were a few subsequent wins, but it wasn't long before she had lost all the money she had with her. "I am DETERMINED to retrieve my losses," she said. A trip to the money-changing office to cash in some bonds at a high commission, just steps from the casino, followed. And again, she lost at roulette. By this time she had lost fifteen thousand rubles. Without a kopeck (one-hundredth of a ruble) left to pay her hotel bill or her train back to Moscow, she ordered Alexis Ivanovitch to go back to the money-changing office to get her last two notes changed. He did so, thinking that she would return to Moscow. Instead, she asked Alexis Ivanovitch to come with her back to the casino. He refused, so the Grandmother went with her old servant Potapitch back to the roulette table and lost almost everything. She sat at the tables for seven or eight hours at a stretch, winning for brief periods led on by false hopes, but mostly losing. Alexis Ivanovitch tells us that the next day she lost all she possessed. All told, she had lost over a 105,000 rubles (approximately 170 pounds of pure gold), a very large sum indeed.

Was it greed that brought her back to the gambling table? Winning eight thousand rubles was a good take, but something sucked her in for more. Excitement? Stardom? Hedonism? Masochism? The Pavlovian call? Greed? Surely Dostoyevsky had something in mind for the character's psyche and her compulsion.

Indeed, the argument for excitement is a compelling one. Who has not felt electricity in the air during a tight horse race, or the intensity of bases loaded with the winning run in the last inning of a World Series, or the thrill of waiting for a tiny steel ball to fall into a pocket of a roulette wheel? One can only imagine the ecstasy, pain, and anxiety of some high-rolling gambler with a wager at any one of those nail-biting moments. Greed seems to be the driving force, but there is also the magic of the game, the enchantment that allures the innocent. Or perhaps, as a good friend of mine once said, "it is the moment when you know you are alive."[1]

That enchantment came slowly to Dostoyevsky's *Gambler* Alexis Ivanovitch, who tells us that when he entered the gaming rooms for the first time in his life he felt oppressed with a need to leave at once, that he could feel his heart heavily, that everything about roulette seemed morally mean and abominable, but that once in the place, radical changes took place to give him promising thoughts of winning at roulette. He sees the folly of the hope and restrains himself at first to simply observe the hungry, restless aspirants in the crowd. Yet there is something terribly alluring about casinos and especially roulette that makes the innocent bystander engage, and, before the spectator comes to his senses, he or she is engulfed in a whirlwind of excitement and illusions of the excitement of life and luck that gaming rooms tend to provide. The eighteenth-century mathematician Pierre-Simon de Laplace attributed such illusions to a false sense of probabilities through errors in judgment over random events, creating wishful biases in favor of the gambler's fallacy.[2]

Gambling is a subject of risk and adventure. Whether it is cards, craps, pool, or roulette, the gambler not only puts him- or herself at risk for losing an immoderate sum, he or she confronts Fortune's grip on the future. The person who drops fifty dollars at a local 7-Eleven for a string of Megabucks tickets is vulnerable to a continuing effort to win the jackpot, especially if he or she strikes a small win. Card players have heard high-rolling stories of Nick the Greek (who once won half a million dollars in a single game) or Titanic Thompson (who won twenty thousand on the turn of a single card), or the MIT group of card-counting blackjack players who, over a ten-year period, walked away from Las Vegas with millions.

Those famous occasional big winners are terrific for the gambling business. That is why the casino industry was happy to promote Sammy Zabib's big Brazil Slingo winnings at a pocket change $779,000 cost to them. When winning is too rare, recreational gamblers feel unsure. Then, all it takes to revive their trust is one successful gambling story. It can draw in ten thousand suckers. Yet there are others who—misunderstanding the odds as well as the law of large numbers—hate the stories of winners. They see opportunity by exploiting their erroneous feeling that success is due.

Returning to Dostoyevsky's narrator of *The Gambler*, Alexis Ivanovitch, we hear him talking about his hot hands impression. He has almost insane thoughts that his luck is hot. It grips him as a prophesized miracle grips a cleric. He believes that his luck has power over mathematics. In short, he believes he is hot, in the sense of Amos Tversky's *hot hand*, with a feeling that he has a better than 50 percent chance of guessing correctly. So he enters the casino in an impassioned state of hope to sit at the Grandmother's table; and, of course, with a masterful writer's fiction-defying reality, he enters an extraordinary winning streak to find himself stuffing banknotes into his pockets, raking in a hundred thousand florins. The hot hand picture was first suggested by Tversky and the psychologists Thomas Gilovich and Robert Vallone in 1985 with an explanation: People reject randomness and the mathematical expected number of runs because the appearance of long runs in short samples seems too purposeful to be random.[3]

It came from Tversky's analysis of the records of a season and a half of the Philadelphia 76ers along with interviews of basketball statisticians, players, and aficionados of the game. Tversky looked at sequences of pairs of shots by players and found a discrepancy between real randomness and the illusion of skill mixed with the perception of randomness in sports. What he found was that— though skill surely matters in basketball—the number of winning streaks was not greater than the expected number of winning streaks in a random sample of data. Often, explanations of human affairs turn to a rejection of randomness, Tversky claimed in a *New York Times* interview.[4] It is as if there is a *hot hands* phenomenon in the body that is not reflected in the data; that is, you feel hot because you're scoring, not that you're scoring because you feel hot.[5] This hot hands illusion generalizes beyond sports to a broad spectrum of gambling games, and especially to dicey casino games such as craps and roulette where random processes play far greater roles than skill-based games such as basketball.

So what is hot, the player's hands or the sequence of scores? In craps, is it the shooter or the dice? In roulette, is it the wheel or the numbers? Earlier in *The Gambler*, Alexis Ivanovitch refers to roulette when he says,

> It was very curious. Again, for the whole of a day or a morning the red would alternate with the black, but almost without any order and from moment to moment, so that scarcely two consecutive rounds would end upon either the one or the other. Yet, next day, or perhaps, the next evening, the red alone would turn up, and attain a run of over two score, and continue so for quite a length of time—say, for a whole day.[6]

We normally have a hard time accepting sequences of wins and losses for what they are—random processes—and feel compelled to impose some sort of order or meaning. Like wishful biases in favor of the gambler's fallacy—which mistakenly predicts a win after a long streak of losses—Tversky's *hot hand fallacy* predicts that people expect long runs of the same outcome to continue. In the early 1970s Tversky and Kahneman gave a cognitive explanation of the gambler's fallacy this way: Though a person may have a fairly good idea of what the law of large numbers says, his or her perception is that a long run of the same outcome is not representative of random behavior and therefore makes irrational economic judgments and decisions.[7] In other words, by that understanding of the law of large numbers, chance may seem to be a self-correcting process where deviations in one direction encourages deviations in the opposite direction to restore equilibrium.[8] But such long runs do not contradict the law of large numbers and indeed are even statistically expected, so no one should be surprised by long runs of the same outcome.[9] After all, in flipping a coin, many of us—even those who should know better—may easily reject true randomness after witnessing four heads appear in a row. Why should we know better? Because even in the most random sequence possible, in twenty flips the probability that four heads will appear consecutively is very high. When people are asked to design strings of random numbers from 0 to 9, they significantly tend to favor strings with fewer repeats than a truly random string would.[10] Naturally, it is hard to mimic truly random sequences of numbers, because cognition will always interfere with stochastic nature. If I write 313249559334823 as a representative string, some unexplainable thought linked each number to its predecessor. What sort of thought? I cannot safely answer—perhaps the 9 after the two 5s came because I was conscious of having two 5s and felt that a third 5 would not be representative; perhaps my mind wished to stay as far from 5 as possible since there were enough low numbers in the developing string. But I do know that a true—or almost true—random number generator would have listed each number in the sequence with absolute ignorance of its predecessor.

When it comes to roulette, players have different views. Some reject the notion that the wheel is hot and—for the most part—accept its randomness. They do not see a long run of reds as an indication that a red is likely to follow.[11] On the other hand, of those who feel that the wheel has some control over the outcome, some do believe that the wheel is hot when seeing three reds appear in sequence. They believe that another red number is more likely to appear than a black. And then there are those of a different personality who feel that black is due.

From the individual's point of view, it is a matter of controlling randomness, at least siding with the illusion of control.[12] The person feels *hot* after a winning streak. In a skill-packed game like basketball it may seem reasonable to say that a winning streak promotes self-confidence to continue scoring, but one may also argue that such a streak puts enormous pressure on the player to continue with such high expectations and that that enormous pressure builds to the point of overwhelming the initial store of self-confidence. Some researchers have argued that the hot hand has an opposing bias, as if there the gambler has some kind of apparitional stock of luck, a belief that there is a certain fixed amount of luck to draw on. Once spent, the likelihood of winning diminishes.[13] A parallel effect, demonstrated in the laboratory, is that gamblers take more risks after losing than after winning.[14] For roulette this takes on the notion that the bets should be placed on outcomes that are *due*, either because they have not occurred for a while or because it is the way the numbers are turning up.[15]

In one field study, gamblers were asked if there is any skill involved in playing a slot machine. The responses varied according to the frequency of gambling. Most frequent gamblers felt that there is equal chance and skill, whereas most infrequent gamblers felt that the game was a matter of mostly chance.[16] This unrealistic desire of control turns truth into wishful memory—we remember our wins without justification and feel compelled to explain our losses; gamblers recall their gambling in broken sequences, each ending with a win.

♣ CHAPTER 15 ♣

Luck

The Dicey Illusion

> Luck is in the slot machine . . . you gotta
> pick out the machine carefully.
> —*Walter, a slot machine addict at Foxwoods*

Foxwoods, at the Mashantucket Pequot Tribal Nation Reservation in Connecticut, boasts being the third largest casino in the world.[1] It is massive! Imagine: it has 7,200 slot machines, 400 tables offering seventeen different kinds of games, a high-tech sports book for betting on horse and dog races all across the United States, and the world's largest bingo hall with 3,200 seats.[2] Walk into the gaming rooms at any hour of any day and you find yourself in the ringing, whistling, beeping hubbub of several thousand slot machines, their lights cursively blinking dizzying matrices of colors to spin minds into believing that anyone can win thousands for pennies. The seats are filled with retirees who are frantically pushing buttons to watch virtual wheels mock their fortunes. Imagine a Kafkaesque complex of more than a quarter million square feet of gaming space filled with 40,000 people—more gamers than the adult population of a medium-size American city!

Many are in control of their daily entertainment funds, but others are not. Some are losing their savings, some their social security checks; some owe more than their net worth, having to borrow from relatives and friends; and still some are losing cars and houses that they have not yet paid for.

I met one man, Walter, who lost two houses. "I have to get them back," he volunteered to tell me. "It'll be slots or blackjack or roulette. I know I can do it. . . . I've been there before."

His friend Barney had just completed the so-called International Slots Tournament. Imagine seventy-five people sitting at slot machines doing nothing but repeatedly pressing GO buttons as rapidly as they can, changing arms as their hands tire. In one half hour, each contestant would have hit a button between five and six thousand times to be in the running. Each machine was programmed to yield the same odds as any other. Scores were based on amounts won; the winners were ranked according to the virtual prize money tallied by the machines. First prize was $1,500. Barney, a retiree who lives a hundred miles away in Yonkers, New York, won fourth prize, $100.

"I come here on weekends and most Tuesdays and Wednesdays," he said. "Four years ago I won $100,000 on a penny machine and I've been coming back ever since. They take out a good chunk in taxes, but, hey, when you walk away with almost $67,000 in your pocket you feel pretty good."

"How often does that happen?" I asked, anticipating the answer.

"Only once at that size, but I sometimes win a thousand bucks and that makes me pretty happy, too."

"So, in the past four years since your big win, have you spent that $67,000 in gaming or are you still ahead?" I asked, pretending that I didn't already know the answer.

"Oh, I'm still ahead."

Barney wouldn't tell me how much he was ahead because—the way I figured it—like most gamblers, he was forgetting most of his losses. He seemed to think that the International Slots Tournament was one of the casino's generous free offers. Others in the tournament were first-time slot players who were watching the virtual payouts of the machines. Many machines seemed to pay out tens of thousands of dollars in 6,000 hits at 25 cents a play. Those novice players were computing good payouts for the cost of playing, not realizing that not all machines were that good at payouts. Smart move for Foxwoods. Very generous.

At one blackjack table there was a veteran player smoking a large cigar behind four tall stacks of $25 chips. Seated next to him was an octogenarian in a wheelchair breathing through an oxygen mask.[3] He was losing. A young waitress in a very short skirt and black tights came to offer him his complementary drinks. He may or may not have been aware that the house had been tallying wins and losses at the table; the generosity of drinks is tied by formula to each player's compounded win-loss score.

Foxwoods is a carpeted gaming mall, not unlike the most upscale American shopping mall lobbies without the escalators and shops, though some sections of the extensive development do house shops and food courts. Ornamented walls and ceilings appear expensive yet graceless, far from the elegance of Monte Carlo where diamonds are forever dazzling, martinis are ordered stirred and not shaken, and Ferraris at curbside are won or lost at the turn of a card in the particularly private rooms by Saudi princes, movie stars, and some of the highest ranks of affluence. Even at the tacky, worn-out Casino Ruhl on the famous Promenade des Anglais in Nice, with its modest dress code, the hot drinks are served in China cups and saucers and the cold in stylish glassware. At Foxwoods, the drinks are served as if takeouts from a Starbucks in a mall food court—the cold in plastic, hot in paper.

Four hours later I returned to see how Barney was doing. I found him at a Bally Bonus 7s, a five-reel machine with several mixed sevens symbols that paid highest on identical sevens. He had won another $100. When I asked him how much it cost him he reached into his pocket and pulled out a red spiral notepad.

"I have it all written down here," he said. "I came with $250 and I now have $230. Well, for twenty bucks I'm sure having a good time. And who knows? The day is still young!"

When I asked if he knew the odds of any of those machines he said he only knew the average odds of the Foxwoods machines and that they had a 91 percent payback.

"Well," I said, with a risk of sounding snide, "doesn't that mean that the casino will give you 91 cents for every dollar you give them?"

"Not at all," he scoffed. "Yeah, they have a 9 percent advantage over me, but that's pretty close to even odds."

He thought he was teaching me a thing about odds, so I couldn't resist asking, "Ever heard of the law of large numbers?"

"That some kinda betting scheme?"

Before we went our separate ways he volunteered a tip: "Hey, let me give you some advice. Scout around the slots for someone who has been playing for a while and not winning. When he leaves, take his seat. That way you'll be at a machine that is due to win something."

Like an amateur, he gambles on his law of averages due.

Nothing is entirely bad or good. The law of large numbers—partly to blame for many gambling problems when misconstrued—is an impressive catch that binds mathematical theory to physical phenomena. It is also responsible for Nature's entropic ways of bringing matter and energy disorder to inert uniformity—a magnificent law responsible for many wonders of our fantastic universe. Who would believe that so many of the vast outcomes of the universe are merely results of colossal successions of dice throws and coin flips?

Accidents happen, DNA replication blunders, lives take strange turns on their short paths, butterflies flaps their wings to cause who knows what. It's all a roll of bones, a casting of lots.

Don't believe it? Watch how liquids diffuse. Pour a small amount of a dissolvable substance such as finely ground cereal (Postum or permanganate) on the left side of a tank filled with water. In figure 15.1, the dots indicate the substance as its concentration diminishes from left to right. Wait a few seconds to see what will happen. The substance will spread from left to right, from higher concentration to lower until it is uniformly distributed throughout the tank.

You might think that there is some force that is driving this tendency of molecules to move from the more crowded region to the less. But there is no such force. There is no more a force that drives this than there is a force for evolution to drive organisms to become fitter. Every molecule in this system is independent of all the others. Every molecule is being knocked about by impact with the water molecules and thereby thumped in an entirely unpredictable direction.

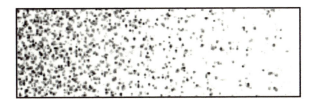

FIGURE 15.1. Diffusion of molecules.

The path of any dissolved molecule is determined randomly. To understand what is happening, place an imaginary line across the tank and ask how probable is it that a molecule on the imaginary line will move toward the right. The answer is that it is equally likely that it will move to the right as to the left. It is as if each molecule flips a coin to decide whether to go left or right. If you take this to be true, then you must agree that more molecules will move from left to right than from right to left, simply because there are more molecules coming from the left side of the imaginary wall than from the right. So the diffusion toward uniformity occurs simply because there is an equal likelihood that the molecules will move in any direction.

The second law of thermodynamics tells us you can play the same game with gasses. Similarly, fleas in a box will quickly disperse in space by jumping in completely random directions without space objectives and without purpose, for even if they found themselves in an almost empty space they will jump again aimlessly in a new random direction.

Very many complex phenomena of nature may be simply explained as flipping a coin or randomly picking a number a gazillion times. And from that huge volume of random numbers chance creates an ever-evolving dynamic world, where permanganate diffuses, gas homogenizes, fleas coincidentally spread out, and compulsive gamblers go bankrupt.

The stock market is a colossally busy assemblage of buyers and sellers. In a typical day, the world market volume of trade may be as high as a billion shares. True, many shares are bought or sold with some knowledge of their future values, but most are bought or sold under speculative guesses based on trends and handicap information no better than what one gets from the Daily Racing Form on Hialeah. Each publicly traded company is tied to hundreds of small ones. Each is linked to others, through competition, shared resources, or possibly dynamic dependence on public consumerism. Place your bets, ladies and gentlemen . . . place your bets. You bet on Microsoft, Citibank, perhaps General Motors. The combined wealth of those three companies would be well over a trillion dollars, some tied up in assets and infrastructure, some in stock loans, and some in investments. Presumably, the shareholders own that wealth.

The daily fluctuations in value of any single company affect those of a pack of others; how could one handicapper know what will happen in a world of daily political, social, or economic events. Hurricanes pass by offshore oil rigs, autoworkers strike to hold onto their benefits, juries grant huge class action damage awards against pharmaceutical firms, orange groves freeze, CEOs are indicted for fraud, and so on. Meanwhile the hedge funds short sell and play the futures market like nineteenth-century rakes at the thermal baths of Baden-Baden.

If a giant goes down—like Pan American or TWA in the 1980s or Bear Stearns and Lehman Brothers in 2008—and the government does not bail it out, all the money those stockholders have invested is raked into an obliterating green hole like late-night chips on a Vegas roulette table.

Though the universe seems to change in random ways, there is a reasonably predictable, and likely inescapable, long-term effect of compounding outcomes. The tenderfoot gambler should understand that the long-term effects of his or her wagers are as inevitable as Nature's entropic flow of matter and energy. In the long run the chips will drift uniformly in the direction of the house's baited treasury.

There will be suckers whenever the Monte Carlo fallacy muddles the law of large numbers. That Monte Carlo fallacy is only one of many false intuitions adopted in gambling. Americans spend billions of dollars on lotteries every year, guided by a mistaken belief that comes from three misguided rules: one, someone has to win; two, I have as good a chance as anyone else; and three, I really think my chances are pretty good. Plausible cause is twisted into feelings of truth. Better-than-even odds evokes the words *most of the time*. *Most* does not mean 100 percent. Nor does it mean more than 60 percent. *Most of the time* simply means a majority of the time, more than 50 percent of the time. It's a useful exaggeration tool, used either when we really don't know the facts or when we purposely want to mislead someone into imagining a percentage much closer to 99 than to 51.

A pair of dice lands on boxcars. Why? Not because there is a 1-in-36 chance and it was time for boxcars to show. The 1-in-36 chance is just a mathematical reason, a brilliant macro-model designed to predict an event influenced by that infamous butterfly. We are so accustomed to thinking in terms of classical mechanics, of cause and effect, that we believe that there should be an undetectable moment—when the dice are held in one of a huge number of possible positions in the palm of a hand, or while spinning in two directions on two axes through the air, continuously affected by countless currents, or striking a surface and bouncing again and again—that inevitably determines boxcars. The logic we use to predict positions of the planets or where a rocket will fall does not apply to the mechanics of small things. We can safely say where the dice will fall—if the toss is not too wild—not how. And we will never know the moment when the dice decide to show boxcars. We *can* say that if the dice are thrown thirty-six times, there is a better-than-even chance that they will show boxcars once. Like all good questions, the one Chevalier de Méré asked in the winter of 1654 didn't end simply with an answer. We now know that it takes twenty-five throws to have favorable odds of throwing boxcars, but a whole new way of thinking about prediction in our enormously complex world of indeterminate events came to us by way of the kinds of questions de Méré's posed.

We have seen that what happens by chance may also tend toward mathematical order and pattern. What appears to be chance may also be simply a matter of counting combinations. Recall from chapter 7 the balls randomly dropping down the Galton board, bouncing from peg to peg left and right with equal likelihood, yet tending to accumulate in a pattern. That pattern was established by mathematical expectation. One could design boards so balls bounce with unequal likelihood from peg to peg and predict the pattern of their accumulation. Pascal and Fermat could not have imagined matter as huge numbers of atoms and molecules crashing into each other at great speeds, bouncing in confusing directions with no apparent pattern. They could not have imagined heredity as the transmission of complex strings of messages written with just four letters and evolution due to random errors in copying those messages. They were thinking of a pair of dice and the smallest number of times a person must throw them to have a better-than-even chance of getting a double six. It would seem that the throwing of dice should have nothing to do with predicting the weather, which might be caused by a pair of butterfly wings; yet, if we could see the infinitesimal connections of accumulated changes caused by those wings, we might look to the butterfly to predict the weather and bring us luck.

Life is a vast sequence of happenings and choices; happenings just happen, but free-will choices are gambles with what's ahead, from which berries to pick in the wild to selecting a college. Analysts have argued that gambling games are representations of inner conflicts, forces separating principles of behavior—*passion* alongside *apathy*, *defiance* against *compliance, good* in opposition to *evil.* I would argue that some—if not most—gambling behavior is primarily connected to an intrinsic desire to manipulate luck in order to validate life, to test the forces of uncertainty under a fantasy of knowing something unknowable or to experiment with the new. Making choices based on scant knowledge is an essential function of consciousness. It is eating from the tree of life. And though Adam didn't know the consequences of biting into something so forbidden and delicious, he was being human, gambling on a choice and testing limits with faith in luck. Gambling is confirmation that *someone* is in control; it is as natural as belief in God. The problem comes when luck is connected to that confused understanding of the law of large numbers that compels the fancy of control, an illusion that one's turn is due. So belief in luck turns out to be as natural as religion. What could be more humanly natural than to have a desire to control one's destiny? It fits the desire to believe that there is some merciful being who will oversee existence, protection, and posterity.

Acknowledgments

I had the help and support of many individuals in writing this book, partly from encouraging and informative conversations, partly from interviews, and partly from constructive suggestions and essential corrections. First, I thank my wife, Jennifer, who read the complete manuscript several times making editorial as well as practical and structural suggestions. She is my greatest support, anchor, and inspiration.

I am exceedingly grateful to the Rockefeller Foundation for an idyllic residency at the Villa Serbelloni in Bellagio, which provided everything a writer could wish for to finish a book: magnificent surroundings, curious and inspiring colleagues, and generous service to bring one to the peak of productivity. In particular, I thank Elena Ongania, Pilar Palacia, and Rubin Puentes for making my stay at the Villa so productive and extraordinarily enjoyable.

With enormous appreciation, to my wonderful colleagues at the Villa Serbelloni who took interest in this project and who contributed so many invaluable ideas: James Brown, Courtney Brkic, Richard Bunce, Deane Calhoun, Lewis Cohen, Mauricio Figueiras, Sonia Jabbar, Rudolph Klein, Kannan Krishnan, Philip Levine, Janis Mattox, Judith Pfeiffer, Bobby Previte, Antony Taubman, Du Yaxiong, and Monica Youn.

Just when I thought I was finished with the book, Peter Madonia of the Rockefeller Foundation suggested that I include a section relating the 2008 economic crisis to risky gambling. Chapter 12 is the result of his terrific suggestion. Thank you, Peter.

Very special thanks to my editor, Vickie Kearn, at Princeton University Press, whose insightful suggestions and careful editing significantly improved the book. To Debbie Tegarden, my production editor, who with wit and humor tolerated my extensive revisions to the end. To Carmina Alvarez, Jenn Backer, Anna Pierrehumbert, Jennifer Roth, and Stefani Wexler (the production staff of editorial assistants, copyeditors, and designers), for taking such good care of the production side of this book and making it as beautiful as it appears here.

Very special thanks go to my expert consultants in gambling addictions: Jeffery Derevensky, Seth Eisenberg, Sarah E. Nelson, Keith Whyte, and Vickii Williams.

To Guido Baltussen, Sorina Eftim, Damon Rein, George Sulimirski, James Tober, and Steve Wynn.

To all my wonderfully supportive friends who have listened to my many stories; to Anne Wheelock for devotedly forwarding relevant news material; and to my constant inspirations for executing good work: Tadatoshi Akiba and William Bown.

As always, I am beholden to the enormously competent Marlboro College library staff, Radmila Ballada, James Fein, Kathleen Packard, Aidan Sammis, and Bonnie White, for gracious and expert assistance in finding cross-referenced material and interlibrary loans with such good cheer, often after hours.

To my brother Barry and sisters-in-law Carole Joffe and Gretchen Mazur, for their constant encouragement.

♣ Appendix A ♣
Description of the Games Used in This Book

In almost every one of the games described there are variations and complex betting rules. Descriptions and rules are brief for the sake of simplicity.

atep: An Egyptian game of guessing the number of upheld fingers.

baccarat: A European card game dating to the fifteenth century similar to basset and faro. No skill or strategy is involved. There are several modern variations of the game. Players bet on the hands they are dealt. High score wins, but scoring has a twist. The score is the remainder after dividing the sum of the face values of the cards by 10.

backgammon: This is one of the oldest board games. Two players move pieces around a board according to rolls of dice. The objective is to remove one's checkers from the board before one's opponent can. A player wins by removing all of his pieces from the board. This game does require skill, which is in strategy.

basset: A card game involving a banker and players who win according to the values of the cards dealt. No skill or strategy is involved.

bingo (the card game): Each player is dealt two hands of cards face down. The value of each hand is the sum of the values of each card, where the cards have blackjack values. On a display board, cards are revealed with chances to bet. Those cards displayed are eliminated. The pot is split between the players with the highest and lowest point totals, unless a player loses every card, in which case he wins the entire pot.

blackjack: Sometimes called 21. This card game is of unknown origin but surely dates from before the seventeenth century, as Cervantes writes about it in one of his stories. The aim of the game is to bet that your hand will total closer to, but not greater than, 21 than the dealer's hand. All picture cards count as 10 and an ace can count as either 1 or 11 at the player's choice.

brag: A uniquely British variant of poker.

Brazil Slingo: A slot machine game in which players are given three spins to match numbers on a playing board.

Caribbean stud: This is a variant of standard poker, where the game is played against the house rather than against other players. No bluffing is possible.

chuck-a-luck: Played with three dice in a wire-frame birdcage. Bets are made. The dealer rotates the cage and calls the outcome.

craps: This is a dice game of several variations, depending on whether it is played on or off the street. It is a simplified American derivative of hazard. Players take turns rolling two dice. The player rolling the dice is called *the shooter*. Other players make bets on the shooter's rolls. In essence the game is played like this: First there is the "come-out roll." If the come-out roll is 7 or 11, those players betting on the shooter wins. If the come-out roll is 4, 5, 6, 8, 9, or 10 that number becomes the *point*. The shooter continues to roll until either the point is made or a 7 is rolled. If a 7 is rolled, the players betting on the shooter lose.

draw poker: A poker game where players are dealt a complete hand before the first betting round. The objective is to develop the hand by drawing new cards to replace discarded ones.

E. O. (evens and odds): Two players (one designated as *odds*, the other *evens*) each hold out a fisted hand, and on the count of three, hold out either one or two fingers. If the sum of fingers shown by both players is an even number *evens* wins, otherwise *odds* wins.

euchre: A card game most commonly played with four people in two partnerships with a deck of 24 standard playing cards, which has been very popular in the United States since the nineteenth century. A player names a specific hand, betting that his or her partner will be dealt that hand.

faro: A card game, very popular in eighteenth-century France and England, and in the nineteenth-century American West. It was played with an entire deck and any number of players against a banker. A card of each of the complete 13 face values was placed face up on a table. Each player lays his or her stake on one of the 13 face values on the layout. The dealer then deals two cards per round alternately to two stacks, the first called the *losing stack* and the second the *winning stack*. If the player staked on the card value dealt to the winning stack bet and that value is not dealt to the losing stack, he or she wins.

fly loo: Players sit around a table, each with a sugar cube. A fly is released. A player stakes his or her bet (usually a considerable sum) that the fly will perch on his or her cube.

gin rummy: A card game in which each of two people play with a hand of seven, nine, or eleven cards. A dealer deals one card face up next to the remaining deck. The first player to go takes a card from the deck or the up-facing card. At each turn a player may take one card from either the top of the face-up pile or the top of the deck. If you draw from the deck, you add it to your hand without showing it and discard a card from your hand to the turned-up pile. The object of the game is to collect a hand where most or all of the cards can be combined into sets and runs and the point value of the remaining unmatched cards is low.

grand hazard: Similar to chuck-a-luck. Players place their bets on the outcome of the three dice that are rolled by the house.

hazard: An ancient English game played with a pair of dice originating from the time of the crusades in the twelfth century. Immensely popular in Europe during the seventeenth and eighteenth centuries.

high/low: Another variant of poker in which the high hand wins half the pot and the low hand wins the other half.

immies: A marble game in which players shoot marbles toward a cardboard box with holes cut out for winning targets.

keno: Ping pong–like balls marked 1 through 80 are blown around in a glass chamber. A "caller" presses a lever opening a tube, where the balls lift one at a time into a "V" shaped tube. The caller records each of 20 balls drawn and a computer tabulates all wagers based on the numbers drawn.

loo: A popular card game played mostly in England from the seventeenth to the nineteenth centuries. Three or five cards are dealt; highest score wins. Players choose to either stay in or drop out. Staying in takes a proportionate share of the pot for each hand they win and pays in an amount equal to the whole pot if they drop out.

mayores: A simple dice game played with three dice—highest roll wins.

monte: This simple game is usually played with just three cards. A dealer places three cards face down on a table, shows one card (the target), and then rearranges them quickly to confuse the player. If the player identifies which card is the target he or she wins the bet. Generally, this is a con game, in which the dealer uses quick sleight of hand.

numbers: An illegal lottery wherein the player bets through a bookie on three or four digits to match those that will be displayed in a newspaper the following day. The winning number is generally the last three numbers in the published daily volume of the New York Stock Exchange.

piquet: A poker-like card game played with a 52-card deck with all the 2–6 value cards removed. The player scoring the most points wins.

rouge et noir: See trente-et-quarante.

stud poker: A poker game where players receive a mix of face-down and face-up cards dealt in multiple betting rounds.

tau: An ancient Egyptian board game with an objective similar to that of backgammon.

Texas hold'em: A poker game where the objective is not winning every hand but deciding when and how much to bet, raise, call, or fold. The strategy is to maximize long term winnings by maximizing expected value on each round.

three-card poker: A fast-moving poker game using only three cards. After seeing the cards dealt, the player can fold and lose the ante bet or raise by placing a bet of equal money. If he chooses to continue, there are three possibilities. The first is that the dealer may not have the required hand of a queen high or higher, in which case the ante is paid out as even money. If the dealer does have a queen high or higher, the player wins if his hand is of higher value than the dealer's, and gets paid out even money on both his ante and play bets. If the dealer's hand is of higher value, the dealer takes the ante and play bets.

trente-et-quarante (also called rouge et noir): Six packs of 52 cards each are used. In this game, a croupier deals out two separate rows of cards (red and black), sequentially. Players bet on either row (with cards counted at face value and face cards counted as 10). When the rows total greater than 30, the hand is settled and whichever row comes closest to 31 wins.

video poker: Players place bets by inserting either money or a bar-coded paper ticket into the machine. A virtual hand is displayed. The player may then press a "Deal" button to draw cards. There are buttons to keep or discard one or more of the cards in exchange for new cards from a virtual deck. After the draw, the machine evaluates the hand and offers a payout if the hand matches one of the hands determined by the machine.

whist: Four players in two fixed partnerships play from a 52-card deck. The cards in each suit are ranked highest to lowest. Each player is dealt 13 cards ranking highest to lowest, in reverse of normal poker. The players must play the same suit as the card that was just played. If a trump suit card is played, then the highest trump (the card determined by the dealer's last card) wins. However, if no trump card is played, then the high card of the suit wins.

Glossary of Gambling Terms Used in This Book

aggie: A marble made of agate.

box: In backgammon chouette, the person who plays alone against all the others.

breakage: The rounding out of the last cent digit on a $2 pari-mutuel race ticket.

bucket shop: A small-time betting shop, often within a coffeehouse, drugstore, or hotel.

bust: In blackjack, what occurs when a player's total card count is greater than 21.

call: To put into the pot the amount equal to the last bet.

captain: In backgammon chouette, the leader of the team playing against the box. He rolls the dice and makes the final decisions for the team.

card counting: Tabulating odds at a casino card game based on what cards have already been dealt.

chouette: Backgammon for three or more players. One player, the box, plays against all the others, who form a team led by a captain.

crap out: In craps, getting a 2, 3, or 12 on the first throw.

credit-default swap: This is a hedge against loss. It is an insurance against an event happening. The buyer pays a premium and in return receives a payout if the event does occur.

crew: In backgammon chouette, members of the team who play with the captain against the box.

croupier: An employee of a casino who deals the game and collects and pays winning and losing bets at table games.

dealer's up card: In gin rummy the first card that is turned up after all hands are dealt. Also any card that is facing upward in a discarded pile.

draw: To take cards from a stack or from the dealer.

drop: Money used to buy chips at a casino.

dunner: A ruffian paid by the gaming house.

flasher: A person who spreads false rumors about how much the house is losing.

handicap: Assigning a bet weighting to racers.

hard hand: In blackjack, a hand that does not include an ace.

hedge: A betting strategy by which a player bets on two opposing outcomes.

hedge fund: Any investment fund that offsets potential losses by hedging investments using a variety of methods including short selling.

hot hand: The belief that a long run of a lucky outcome will continue.

ice: Money paid to police and officials for permission to gamble illegally.

meld: To declare or display a card or combination of cards in a hand.

morning-line odds: A handicapper's guess of the odds on a race to be run in the afternoon.

ninja loan: A type of subprime loan issued to borrowers with *no income, no job, and no assets.*

odds: The number of unfavorable events to the number of favorable.

pari-mutuel betting: A betting system in which bets are pooled. Taxes and a house take are subtracted, and payoff is shared among all winning bets.

pip: A dot indicating a unit of numerical value on dice, dominoes, or playing cards.

place: Coming in no worse than second in a race.

point: A number or total in a gambling game on which a bet can be placed.

post time: Starting time of a race.

pot: Money at stake in a betting scheme that is contributed by players.

public handicapper: A journalist working for a racing sheet who predicts winners.

puff: A person who loans seed money to hook unsuspecting novices.

punter: An amateur gambler.

raise: To put more money into the pot than any previous player.

rake: A person who wastes fortunes and incurs debts.

selling long: Selling a commodity, financial instrument, or security that the seller owns.

selling short: Selling a commodity, financial instrument, or security that the seller does not own (but borrowed) at the time of the sale.

sharp: A smart cheater.

shooter (in dice game): The player who tosses the dice.

shooter (in marbles): A marble used to knock off other marbles in a game of marbles.

show: Coming in no worse than third in a race.

Shylock: A person who loans seed money to hook unsuspecting novices.

soft hand: In blackjack, a hand that includes an ace.

stand: In blackjack, to refuse to draw another card.

sub-prime: A financial term referring to borrowers who do not meet prime underwriting guidelines.

tote (totalizer): The display that totals the number of bets and approximate odds at racing tacks while betting windows are open.

trick: A scoring card in a card game. Also refers to a genre of card games where play goes through a series of rounds, called tricks. The object of those games is to win the tricks, or the cards played in taken tricks.

vigorish: The percentage taken by the house.

welsh: To go back on a promise to pay a debt.

Let p be the mathematical probability of success and k/N be the actual success ratio. The question is how close is k/N to p? In modern notation, where ε represents any small positive number chosen,

$$P\left[\left|\frac{k}{N} - p\right| < \varepsilon\right]$$

converges to 1, as N grows large.

Using Chebyshev's inequality and some simple algebra, we may prove the theorem. If we perform a binomial experiment N times, then (by definition of the probability) the expected ratio of successes is k/N, and the standard deviation of the ratio of successes (again by definition) is

$$\sigma = \sqrt{\frac{pq}{N}},$$

where q = 1 − p. Now let

$$h = \frac{\sqrt{N}}{L}$$

where L is some number we will choose later. We have

$$\frac{1}{h^2} = \frac{L^2}{N},$$

and so

$$h\sigma = \frac{\sqrt{pq}}{L}.$$

By Chebyshev's inequality, the probability is smaller than (or equal to) L^2/N that the ratio of success, k/N, differs from p by

$$\frac{\sqrt{pq}}{L}$$

or more. Expressed more succinctly, this says:

$$P\left(\left|\frac{k}{N} - p\right| \geq \frac{\sqrt{pq}}{L}\right) \leq \frac{L^2}{N}.$$

So if L^2/N and

$$\frac{\sqrt{pq}}{L}$$

are both small, then our last sentence would translate as: the likelihood is small that the ratio of successes differs by more than a very small amount. Now here is the dazzling trick. Remember that we can choose L to be any number we wish. Choose L so that

$$\frac{\sqrt{pq}}{L}$$

is small. (Since p and q are already fixed, we can do that.) Then choose N large enough so that L^2/N is as small as we need. This proves the weak law, for it says that for any ε and δ there is a large N, for which

$$P\left(\left|\frac{k}{N} - p\right| \geq \varepsilon\right) \leq \delta$$

holds, and that is precisely the weak law of large numbers.

Binomial frequency curve

Suppose we have a game with only two outcomes: success and failure. If the probability of success is p and the probability of failure is q, then we know that $p + q = 1$. Then suppose the game is played n times. There are two possible outcomes on each trial. It is possible for a player to win a few of those n trials and lose a few. If we let k be the number of wins that occur in n trials, we know that the number of different ways those wins can occur is C_k^n (see combinations). The probability of k successes in n plays is given by $P(k\ successes\ in\ n\ tries) = C_k^n p^k q^{n-k}$. For example, if $n = 4$, there is only one way for the game to be won four times, by winning on every play, but there are six ways to win twice—win-win-lose-lose, win-lose-win-lose, win-lose-lose-win, lose-win-win-lose, lose-win-lose-win, and lose-lose-win-win. If $p = 1/4$, then

$$P\left(0\ successes\ in\ 4\ tries\right) = C_0^4 p^0 q^{4-0} = 1\left(\frac{1}{4}\right)^0 \left(\frac{3}{4}\right)^4 \approx 0.32$$

$$P\left(1\ success\ in\ 4\ tries\right) = C_1^4 p^1 q^{4-1} = 4\left(\frac{1}{4}\right)^1 \left(\frac{3}{4}\right)^3 \approx 0.42$$

$$P\left(2\ successes\ in\ 4\ tries\right) = C_2^4 p^2 q^{4-2} = 6\left(\frac{1}{4}\right)^2 \left(\frac{3}{4}\right)^2 \approx 0.21$$

$$P\left(3 \text{ successes in } 4 \text{ tries}\right) = C_3^4 p^3 q^{4-3} = 4\left(\frac{1}{4}\right)^3\left(\frac{3}{4}\right)^1 \approx 0.047$$

$$P\left(4 \text{ successes in } 4 \text{ tries}\right) = C_4^4 p^4 q^{4-4} = 1\left(\frac{1}{4}\right)^4\left(\frac{3}{4}\right)^0 \approx 0.004.$$

As the number of trials increases, the bar graph of these probabilities begins to look like a curve. That curve is what is called the *binomial frequency curve*.

Combinations

We distinguish between permutations and combinations: The number of permutations of *n* things taken *r* at a time is the number of ways *r* things can be sequenced in a set of *n* things. For example, we might ask how many different ways can *n* balls be grouped, if each group must end up containing exactly *r* balls (assuming *r* is less than *n*)? There are *n* ways to choose the first ball, $(n-1)$ ways with the second, and so forth. However, the choices end at choosing ball number $(n-r+1)$. So there are

$$n \cdot (n-1) \cdot (n-2) \dots \cdot (n-r+1) = \frac{n!}{(n-r)!}$$

ways of arranging *n* balls taken *r* at a time.

The number of combinations of *n* things taken *r* at a time is defined to be the number of subsets that can be derived from a given set, regardless of permutation sequence. For example, consider a box with three balls marked **A, B,** and **C.**

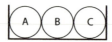

If we wish to pick out two balls from the box without regard to the order in which the balls are picked, we would have three possible combinations.

However, if we cared about order, that is, considering picking **A** first and **B** second as different from picking **B** first and **A** second, then we would have 6 possible combinations.

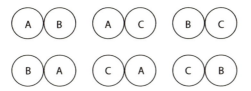

In general, the number of combinations of n things taken r at a time is given by

$$\frac{n!}{r!(n-r)!}.$$

To see this we play a little trick. First, we use the symbol C_r^n for the number of combinations of n things taken r at a time. For each one of those combinations we permute to r objects. There are $r!$ such permutations for each combination and hence a total of $r!C_r^n$ combinations of the permuted objects. That total is equal to the number of permutations of n things taken r at a time, which we already know is equal to

$$\frac{n!}{(n-r)!}.$$

Hence, C_r^n, the number of combinations of n things taken r at a time, is equal to

$$\frac{n!}{r!(n-r)!}.$$

More specifically, if we have a box containing five letters A, B, C, D, and E, and we wish to blindly pick out two letters, in how many different ways can that be done? It is easy to make a list of all the possible combinations of five letters taken two at a time: AB, AC, AD, AE, BC, BD, BE, CD, CE, DE. There are only ten possibilities, just as C_2^5 predicts.

$$C_2^5 = \frac{5!}{2!(5-2)!} = \frac{5!}{2!3!} = \frac{1\cdot2\cdot3\cdot4\cdot5}{(1\cdot2)(1\cdot2\cdot3)} = \frac{4\cdot5}{1\cdot2} = \frac{20}{2} = 10.$$

In general, the symbol C_r^n stands for the number of ways of arranging combinations of n objects taken k at a time, which turns out to be

$$\frac{n!}{k!(n-k)!}.$$

The symbol $k!$ stands for the product of all numbers from 1 to k. After three hundred years of use, the notation for n things taken k at a time has not yet standardized. Even today, we find it symbolized by

$$C_k^n, \; C_{n,k}, \; {}_nC_k, \; C(n,k), \text{ or } \binom{n}{k}.$$

Expected value (sometimes called expectation)

This is a term that is used extensively in this book. In any single betting game there are payoffs for outcomes. The general definition of expected value is more than what we need in this book. First, we assume that no two outcomes can occur simultaneously. Here it is defined as a weighted sum, the sum of all possible values of a payoff, each multiplied by its probability of occurrence. For example, suppose you are offered \$20 for rolling a 4 with a pair of dice and \$360 for rolling a 2. The probability of rolling a 4 is 1/12 and the probability of rolling a 2 is 1/36. The expected value is then

$$\left(\frac{1}{2}\right)(\$20) + \left(\frac{1}{36}\right)(\$360) = \$20.$$

This example does not take the entrance fee into account. Here's another example. A \$100 bet on red in American roulette (where there are 0 and 00 in addition to 36 positive numbers) has an expected value computed as

$$\left(\frac{18}{38}\right)(\$100) + \left(\frac{20}{38}\right)(-\$100) = \frac{-\$200}{38} = -\$5.26.$$

Mean

The mean as used in this book is defined to be the average of a distribution of values, that is, the quotient of the sum of the values divided by the number of values. For example, the mean of the five values 3, 6, 2, 10, and 14 is

$$\frac{3 + 6 + 2 + 10 + 14}{5} = 7.$$

Odds

When we say the odds are 50 to 1 against, we mean that there are 50 ways to lose, just 1 way to win, and that each way has an equal chance. More generally, the odds against an event are computed from the ratio

$$\frac{1 - p}{p},$$

where p is the probability of the event occurring. For example, if $p = 1/3$, the ratio

$$\frac{1 - p}{p}$$

reduces to the ratio to 2/1, and we say that "the odds are 2 to 1 against the event happening." Likewise, the odds in favor of an event are computed from the inverse ratio

$$\frac{p}{1 - p}.$$

To compute the probability p from known odds a to b against the event happening, we use the formula

$$\frac{b}{a + b} = p.$$

Odds are used rather than probabilities in gaming because it is easier to compute winnings; a winning bet of \$1 paying 2 to 1 would collect \$2, an amount that already includes the original stake. Note, though, that the bookmaker will also have to take a commission. Even odds or even money means 1-to-1 odds.

Probability

Literally, probability is a measure, on a scale from 0 to 1, of an event's likelihood—0 being impossible, 1 being certain. However, there are two ways to make that measurement. One is to take a look at the relative frequencies from a large sample. So, if in some hospital 1,000 babies are born each year and on average 480 are girls, then one might be persuaded that the probability of a girl being born is 480/1,000, or 0.48. Another way to measure likeliness—when convenient and appropriate—is to simply take the ratio of the number of favorable outcomes over the number of possible outcomes. For example, there are 6 possible outcomes in the rolling of a fair die and only 1 possible way of rolling a three. So the probability of getting a three is 1/6.

We use the symbol $P(A)$ to indicate the probability that event A will occur. For independent events A and B, where the occurrence of A is not related to the probability of the occurrence of B, the probability of A *or* B happening is the product of the probability of A and the probability of B. So, if A and B are possible outcomes, the probability of A *and* B happening is the product $P(A) \times P(B)$. If A and B cannot possibly occur simultaneously, then the probability of A *or* B happening is the sum $P(A) + P(B)$. For example, *P(red queen and queen of spades)* $= 1/26 + 1/52$.

We often use the symbol p to denote a probability that a certain event will successfully occur, as long as the event in question is understood. In those cases, the symbol q would denote the probability that the event would not occur. Therefore $q = 1 - p$.

Standard deviation

The standard deviation σ is a measure of how scattered the outcomes are from the mean as well as how quickly the distribution curve spreads from the mean. In a sample of n numerical data values x_i, it is computed as

$$\sigma = \sqrt{\frac{\left(x_1 - \mu\right)^2 + \left(x_2 - \mu\right)^2 + \ldots + \left(x_n - \mu\right)^2}{n}}.$$

There are other ways of computing standard deviation. If the numerical data consist of a distribution of n observations of success, where the probability of success p is known, then $\sigma = \sqrt{npq}$.

When applied to the data under a normal curve, 68 percent of the area lies between 1 standard deviation from the mean, and 95 percent of the area lies between 2 standard deviations from the mean.

Standard normal curve

The graph of the function,

$$y = \frac{1}{\sqrt{2\pi}} e^{\frac{-x^2}{2}},$$

where e is the base of the natural logarithm, approximately 2.1718, is called the *standard normal curve*. (See figure 8.5.) The important thing to know is that every binomial frequency curve may be transformed (through shifting and scaling) into the standard normal curve and used as an approximate model for many natural phenomena resulting from chance behavior.

♣ Appendix E ♣

Callouts

Chapter 1

Callout #1

Tau goes back to at least 1600 BCE—Queen Hatasu's set is on exhibit at the British Museum. In the Egyptian gallery in a case containing the throne of Queen Hatasu there is a checkers board together with its lion's-headed pieces of light and dark wood. The Etruscan room of the British Museum also has several fine examples of onyx, porcelain, stone, and terra cotta astragals.

Callout #2

With the broken bar represented by 0 and the unbroken represented by 1 in binary expansion we get numbers 0–7 as follows:

$$0 \cdot 2^2 + 0 \cdot 2^1 + 0 \cdot 2^0 = 0 + 0 + 0 = 0$$

$$0 \cdot 2^2 + 0 \cdot 2^1 + 1 \cdot 2^0 = 0 + 0 + 1 = 1$$

$$0 \cdot 2^2 + 1 \cdot 2^1 + 0 \cdot 2^0 = 0 + 2 + 0 = 2$$

$$0 \cdot 2^2 + 1 \cdot 2^1 + 1 \cdot 2^0 = 0 + 2 + 1 = 3$$

$$1 \cdot 2^2 + 0 \cdot 2^1 + 0 \cdot 2^0 = 4 + 0 + 0 = 4$$

$$1 \cdot 2^2 + 0 \cdot 2^1 + 1 \cdot 2^0 = 4 + 0 + 1 = 5$$
$$1 \cdot 2^2 + 1 \cdot 2^1 + 0 \cdot 2^0 = 4 + 2 + 0 = 6$$
$$1 \cdot 2^2 + 1 \cdot 2^1 + 1 \cdot 2^0 = 4 + 2 + 1 = 7$$

Chapter 2

Callout #3

It would have been possible for mathematicians in the thirteenth century to discover the number of permutations (i.e., orderings) of n objects. It is the same as the number of different ways of filling n boxes with n balls. There are n choices for the first ball, $(n-1)$ choices left for the next ball, $(n-2)$ choices for the next until there is only 1 ball left. The last ball has no choice. Hence there are $n \cdot (n-1) \cdot (n-2) \ldots \cdot 1 = n!$ different ways of arranging n balls in n boxes. (We are using the symbol $n!$ to denote the product of all integers from 1 to n.)

Callout #4

There are two laws of large numbers: the strong and the weak. Here we are using what is called the *strong law of large numbers*, which says that with probability 1 (certainty) the sample mean approaches the true mean as the sample size n grows large. The weak law says that (with some probability $P(n)$ that approaches 1 as n grows large) the sample mean approaches the true mean as n grows large.

Callout #5

The old general rule was clever, but not correct. It was assumed that games involving two dice were similar to games with one die. Suppose there is 1 chance in N of winning a game on a single trial. Let k be the number of trials needed to have a better-than-even chance of success. The rule assumes that the ratio k/N is a constant. In the case of dice, it takes 4 throws of a single die to have a better-than-even chance of throwing a 6. Hence $k/N = 4/6$ or $2/3$. For throwing double six with a pair of dice, $N = 36$ and hence $k/N = k/36$. But that ratio must still equal $2/3$, so k, the number of throws needed to have a better-than-even chance of throwing two 6s, must be 24.

Callout #6

The solution is a clever one. In a letter to Fermat dated August 24, 1654, Pascal gives this example: Suppose that two players A and B begin a game of dice in four rounds and for some reason have to abandon the game after three rounds. A special case will simplify the general solution. Suppose that player A is short two points and player B is short three. We are therefore trying to find the number of ways the four points can be distributed between players A and B. Pass to an equivalent problem: They are to flip coins, four at a time—heads player A wins, tails player B wins. Now mark down the sixteen possibilities, each column indicating the possibilities of the flip of four coins.

```
H  H  H  H  H  H  H  H  T  T  T  T  T  T  T  T
H  H  H  H  T  T  T  T  H  H  H  H  T  T  T  T
H  H  T  T  H  H  T  T  H  H  T  T  H  H  T  T
H  T  H  T  H  T  H  T  H  T  H  T  H  T  H  T
```

Since player A is short two points, any column containing two or more Hs indicates a win for A. Any column containing three or more Ts indicates a win for B. From the array, we tally 11 wins for player A and 5 wins for player B. Hence, the fair division of the stake is 11 to 5 in favor of player A. How clever!

Callout #7

Today anyone who understands how to use logarithms can quickly solve the problem by first solving the equation

$$1 - \left(\frac{35}{36}\right)^n = \frac{1}{2}.$$

Simply subtract 1/2 from each side to get

$$\frac{1}{2} - \left(\frac{35}{36}\right)^n = 0 \text{ or } \frac{1}{2} = \left(\frac{35}{36}\right)^n.$$

Take the natural logarithm of both sides to get

$$Ln\left(\frac{1}{2}\right) = Ln\left(\left(\frac{35}{36}\right)^n\right) \text{ or } Ln\left(\frac{1}{2}\right) = nLn\left(\frac{35}{36}\right).$$

Isolate n by dividing both sides by

$$Ln\left(\frac{35}{36}\right) \text{ to get } n = \frac{Ln\left(\frac{1}{2}\right)}{Ln\left(\frac{35}{36}\right)} = 24.6.$$

If n is to be an integer that does the trick, then it must be 25. The odds are against throwing a double six in 24 trials, but in favor of throwing a double six in 25 trials.

$$1 - \left(\frac{35}{36}\right)^{24} < \frac{1}{2}, \text{ but } 1 - \left(\frac{35}{36}\right)^{25} > \frac{1}{2}.$$

Callout #8

There are earlier reports of the triangle starting with the work of the tenth-century Indian mathematician Halaydha, who wrote a commentary on the *Chandas Shastra* (the Sanskrit treatise on the study of poetic meter) where he notices that the diagonals of the triangle sum to what was later called the Fibonacci numbers. Such a triangle may have existed at such an early date; however, it surely did not consider the formula for construction and may simply have listed enough rows to be useful.

If we look at the triangle as arranged in figure AP-E-1, we see that the diagonals do indeed sum to the Fibonacci numbers.

The triangle appeared in the works of the twelfth-century Chinese algebraist Chu Shǐ-kié and later appeared on the title page of Petrus Apianus's *The Arithmetic Book* in 1527 (which appears in the painting *The Ambassadors* [1533] by Hans Holbein the Younger), more than

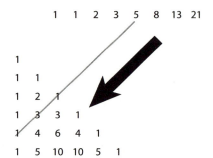

FIGURE AP-E-1. Diagonals of Pascal's
triangle summing to Fibonacci
numbers.

a century before Pascal ever investigated the triangle named after
him. In modern Iran the triangle is known as the Khayyám triangle,
after the famous Persian poet and mathematician Omar Khayyám,
who used this triangle in the twelfth century to compose a method
for finding *n-th* roots. In modern China it is called Yang Hui's triangle in honor of another mathematician who introduced it to China
in the thirteenth century. In Italy it is Tartaglia's triangle, after the
mathematician Niccolò Tartaglia, who lived a century before Pascal.
However, Pascal, a collector of many results that had already been
known about the triangle, broadly used them for probability theory.

Chapter 4

Callout #9

Fly loo is an American gambling game in which sugar cubes are set in
various places on a table and a fly is released in the vicinity in the hopes
that sooner or later the fly will land on one winning cube. The game
is duly described as far back as 1883 as a popular ocean voyage game.

There is also an old card game by the name of loo in which players
are dealt cards and the highest scoring on the single deal wins the
pool. The name seems to come from the word *halloo*, which means
a call to attract attention; presumably, the call is addressed to the
deck. How this translates to fly loo is anyone's guess, but it seems
reasonable to assume that in the course of the game the indecisive
insect is *looed* by each player.

Chapter 7

Callout #10

The probability of k heads is $1/2^k$. However, a run can start at toss 1, or toss $(n-k)$, or any toss in between. So the probability of a run of k heads is:

$$\underbrace{\frac{1}{2^k} + \frac{1}{2^k} + \cdots + \frac{1}{2^k}}_{n-k+1 \text{ times}}$$

Chapter 8

Callout #11

The symbol Σ is simply a shorthand way of saying: add the terms that appear on the right. For example

$$\sum_{k=51}^{100} k$$

means add all the integer numbers between 51 and 100. The answer is 3,775. In our case, we wish to add all the terms

$$C_k^{100} \left(\frac{9}{19}\right)^k \left(\frac{10}{19}\right)^{100-k}$$

where k increases from 51 to 100, i.e.,

$$\sum_{k=51}^{100} C_k^{100} \left(\frac{9}{19}\right)^k \left(\frac{10}{19}\right)^{100-k}$$

is just a more compact way of writing the long and messy sum:

$$C_{51}^{100} \left(\frac{9}{19}\right)^{51} \left(\frac{10}{19}\right)^{49} + C_{52}^{100} \left(\frac{9}{19}\right)^{52} \left(\frac{10}{19}\right)^{48} + C_{53}^{100} \left(\frac{9}{19}\right)^{53} \left(\frac{10}{19}\right)^{47}$$

$$+ \ldots + C_{100}^{100} \left(\frac{9}{19}\right)^{100} \left(\frac{10}{19}\right)^{0}.$$

Callout #12

A qualification is necessary here: Before the second half of the twentieth century, the standard normal distribution was the only practical way to compute the probabilities under a binomial distribution bar graph for an unreasonably large number of bars. Even 100 is too large a number to practically make the calculations by hand. In the computer age, the matter is a simple computation,

$$P(s < x < t) = \sum_{k=s}^{t} C_k^n p^k q^{n-k} .$$

A computer can easily handle $n = 1,000$, whereas the fastest human calculator would have trouble making such calculations in a year. Some modern statistics textbooks have forgotten the real purpose of approximating binomial distributions by the standard normal distribution. It is not to make computations of theory easier but to translate an observational model to a theoretical one in which computations can be made. There are serious practical reasons for using the standard normal distribution to approximate the binomial distribution, but facility with computation is not one of them.

Callout #13

We must interpret what the x and y mean here. The graph of the function $y = e^x$ is a curve that is very small for negative values of x. It passes through the point $(0,1)$ and grows exponentially for positive values of x. A minus sign before the x simply revolves the curve about the line $x = 0$ (the y axis). Squaring the x does two things: It makes the function grow faster and it forces the function to be symmetric about the line $x = 0$ (the y axis). Raising a number to a power that is less than 1 decreases the number. Put all this together and you can see that the function $y = e^{-x^2}$ is a bell-shaped curve. We want a function of the form $y = ae^{-b(x-c)^2}$, where a, b, and c are particular chosen numbers.

   c shifts the curve horizontally to center the mean over the origin
   b shrinks the curve horizontally to transform the units into units
      of concentration

a expands the vertical so the area under the curve and above the x axis remains equal to 1, keeping the curve a real probability distribution curve

This new curve approximates our bar graph distribution, a graph constructed from computations of the binomial $(p+q)^n$. It turns out that for $y = ae^{-b(x-c)^2}$, computing areas is generally hard. However, there is one very big advantage: We have reduced an infinite number of curves to the one curve, the standard normal distribution curve

$$Y = \frac{1}{\sqrt{2\pi}} e^{\frac{-X^2}{2}},$$

and its areas have already been computed for us by tireless, persistent mathematicians.

Callout #14

Figure AP-E-2 best describes the notions of inflection point, concave up and concave down.

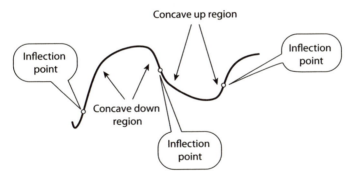

FIGURE AP-E-2. Notions of inflection point.

Callout #15

There is a stronger theorem, the strong law of large numbers, which says that the probability that the sample averages converge to the expected value as the sample size goes to infinity is *almost surely* equal to 1. Note that there is a difference between this strong law and the weak law. The weak law says that the probability can be made as close to certainty as we wish by increasing the number of picks, whereas the strong law says that the probability is *almost surely* equal to 1. Keep in mind that there is a subtle difference between saying that the probability of an event *almost surely* equals 1 and saying that the event is sure to happen. An event is sure to happen if no other event can occur. Here is an example: Suppose you were to randomly and blindly pick points on the real number line. Remember that the cardinality of the real numbers is greater than that of the rationals; for those who know measure theory, the rational numbers are a set of *measure zero*. The probability of hitting an irrational number is equal to 1 and the probability of picking a rational is 0, yet it is possible to pick a rational number.

Callout #16

For a binomial experiment with n trials, the range is between

$$\frac{1}{2}n - \frac{1}{2}s\sqrt{n}$$

and

$$\frac{1}{2}n + \frac{1}{2}s\sqrt{n},$$

where s is the standard deviation. If $s = 1$, there is a 68 percent chance; for $s = 2$, a 95 percent chance; and for $s = 3$, a 99.7 percent chance.

Callout #17

This is a matter of computing the integral

$$\int_{-5/6}^{5/6} \frac{1}{\sqrt{2\pi}} e^{-\frac{x^2}{2}} \, dx \, .$$

Chapter 10

Callout #18

Poker does require skill, but the probability of getting a royal flush is 0.000000154. Think about it this way: There are

$$C_5^{52} = \frac{52!}{(5!)(47!)} = 2{,}598{,}960$$

different possible (five-card) poker hands and just 4 royal flushes. Hence, the probability of a royal flush is

$$\frac{4}{2{,}598{,}960} = 0.00000154 \, .$$

(The odds are 649,739 to 1.)

The probability of drawing one of the 40 possible hands of a straight flush is 0.000139, another near impossibility. The probability of simply getting four-of-a-kind is 0.00024; there are only 624 possible hands. One way to think about it is as follows: Let's start with some arbitrary kind, say, kings. There are 4 kings in a deck and 48 other cards, so kings can be selected in only one way and the other cards in 48 different ways. Hence, there are 48 different ways of selecting 4 kings. There are 13 different kinds of cards and hence $13 \times 48 = 624$ different ways of selecting four-of-a-kind. Therefore, the probability of drawing four-of-a-kind is

$$\frac{624}{2{,}598{,}960} = 0.00024 \, .$$

The odds are 4,164 to 1.

To compute the probability of a full house (two-of-a-kind together with three-of-a-kind), we first note that there are 13 possibilities for the pair and that the pair can be chosen in

$$C_2^4 = \frac{4!}{(2!)(2!)} = 6$$

different ways. The triple can be selected in

$$C_3^4 = \frac{4!}{(3!)(1!)} = 4$$

ways and that particular triple can be chosen in 12 ways (because the value of the pair is no longer available). Hence, there are $13 \times 6 \times 4 \times 12 = 3{,}744$ different ways of drawing a full house and the probability of drawing one is

$$\frac{3{,}744}{2{,}598{,}960} = 0.0014 \, .$$

The odds of that happening are 693 to 1.

There is 1 chance in 46 of being dealt three-of-a-kind. Consider finding the probability of getting just three-of-a-kind (any three cards with the same face value). To draw three-of-a-kind we have:

- 13 kinds of cards, A, 2, 3, . . . J, Q, K, so there are 13 ways to pick one kind;
- 4 suits and therefore 4 ways of picking three suits from 4;
- 66 ways of picking two different kinds from the remaining twelve;
- 4 ways of picking 1 suit from 4, but that must be picked twice.

Therefore, there are $13 \times 4 \times 66 \times 4 \times 4 = 54{,}914$ different ways to get three-of-a-kind. Hence, dividing 54,914 by the number of possible hands, the probability of being dealt three-of-a-kind is 0.0211, and the odds are approximately 46 to 1.

Strangely enough, computing the odds of being dealt two-of-a-kind (a single pair with the other three cards of distinct value) is a bit more complicated. One way is to first note that two kings can be chosen from four kings in

$$C_2^4 = \frac{4!}{(2!)(2!)} = 6$$

different ways. The other three cards are to be chosen from the 12 remaining face values, each card drawn from four cards of each distinct value. So the number of ways the non-pair can be drawn is

$$C_3^{12}(4)(4)(4) = \frac{12!}{(3!)(9!)}(12) = 14{,}080,$$

and therefore the number of distinct hands with a pair of kings is $6 \times 14{,}080 = 84{,}480$. Since the pair of kings could have been a pair of any of 13 different face values, there are $13 \times 84{,}480 = 1{,}098{,}240$ different hands containing two-of-a-kind. Hence, the probability of being dealt such a hand is

$$\frac{1{,}098{,}240}{2{,}598{,}960} = 0.423.$$

The odds are 2 to 1.

Callout #19

Beware: the meaning of *odds,* as opposed to *payoff odds,* is distinguished by context. Payoff odds may be (and often are) different than the odds of winning the game. In the case of the wheel of fortune with 25 numbers, the player may be offered payoff odds of 23 to 1, meaning a payoff of $5.75 on twenty-five cents for every win; however, if the wheel is perfectly balanced, the odds of winning are 24 to 1, on average.

Callout #20

To see that this is true, let $S_n = p + pr + pr^2 + pr^3 + ... + pr^n$, for some finite number n. Then $rS_n = pr + pr^2 + pr^3 + ... + pr^n + pr^{n+1}$. Subtract rS_n from S_n to get $S_n - rS_n = p - pr^{n+1}$. Solve for S_n to get

$$S_n = p\left(\frac{1-r^{n+1}}{1-r}\right).$$

As long as the absolute value of r is less than 1 (which it is in our case), the right side approaches

$$p\left(\frac{1}{1-r}\right)$$

as n approaches infinity and the left approaches $p + pr + pr^2 + pr^3....$

Appendix C

Callout #21

You may be wondering how it is that we are using

$$\sigma = \sqrt{\frac{pq}{N}}$$

when in chapter 8 we said that $\sigma = \sqrt{Npq}$. Normally, in performing a binomial experiment a large number of times the standard deviation is computed as $\sigma = \sqrt{Npq}$, where p is the probability of success, q is the probability of failure, and N is the number of times the experiment is performed. However, in this case we are not interested in k, the number of successes, but rather the success ratio k/N of a succession of N independent Bernoulli trials, (success or failure experiments, all with probability of success p). The distribution of such a large number of sample success ratios is approximated by a normal distribution whose mean is p and standard deviation is

$$\sigma = \sqrt{\frac{pq}{N}} .$$

Notes

FRONTMATTER

1. Abraham De Moivre, *The Doctrine of Chances*, originally published in 1716 and reprinted as a photographic reprint by the American Mathematical Society in 2000.

INTRODUCTION

1. *The Iliad of Homer*, trans. Richard Lattimore, Book XV (New York: Harper Collins, 1974), lines 185–92.
2. From Chaucer, *The Canterbury Tales*, trans. Nevill Coghill (Baltimore: Penguin, 1952), 205.

CHAPTER 1
PITS, PEBBLES, AND BONES

1. The first intact human skull fossils were found in 1856 in the Neander Valley near Düsseldorf, Germany. The word Neanderthal comes from the translation of Neander Valley—the German word for *valley* is *tal*. Neandertal is an alternate spelling.

2. Our typical image of Neanderthals comes from unfortunate folklore depicting the ill-fated man as an ignorant wretched brute living in caves. But, as Stephen Jay Gould has pointed out on many occasions, the caveman's image is simply "an artifact of preservation," and that Neanderthal man had a highly ritualized gentle society.

3. Philip Liebermann, *Uniquely Human: The Evolution of Speech, Thought, and Selfless Behavior* (Cambridge, MA: Harvard University Press), 1993. For an interesting assessment of Liebermann's theory, see Bo Gräslund, *Early Humans and Their World* (London: Routledge, 2005), 109–11.

4. See Dale Guthrie, *The Nature of Paleolithic Art* (Chicago: University of Chicago Press, 2006), vii–x.

5. Antelope astragals were prized for their superiority of randomness.

6. Exod. 28:30.

7. *Pentateuch & Haftorahs*, ed. J. H. Herz (London: Sonchino Press, 1987), 341–42.

8. The British archaeologist Sir Leonard Woolley discovered this game while excavating the Royal Cemetery of Sumerian kings at Ur in Mesopotamia before World War I.

9. Some are housed in the Berlin Museum.

10. Edward Falkener, *Games Ancient and Oriental and How to Play Them* (London: Longmans, Green and Co., 1892), 103–9.

11. John Ashton, *A History of Gambling in England* (1898; rept., Montclair, NJ: Patterson Smith, 1969), 12.

12. The text and illustrations of *Libro de Los Juegos* are reproduced in Alfonso El Sabio's "Libro de Acedrex, Dados e Tablas," *Zeitschrift für Kunstgeschichte* 53:3 (1990): 277–308. The original *Libro de Los Juegos* is in the library of the monastery of San Lerenzo del Escorial, outside Madrid. I have not seen this text but am told that it separates skill from chance. The first part is on chess as an example of a game of pure skill; the second on dice; and the last on what were called *tables*, that is, games of both skill and chance. Another source is Sonja Musser Galladay, "*Los libros de acedrex dados e tablas*: Historical, Artistic and Metaphysical Dimensions of Alfonso X's *Book of Games*" (Ph.D. thesis, University of Arizona, 2007).

13. Though there is no comprehensive definition to distinguish gaming from gambling, I use the word *gaming* to refer to gambling in any nonphysical event where the gambler is an active participant. For example, playing cards, dice, or roulette would be considered gaming, whereas gaming would not refer to betting at a cockfight, horse race, or lottery.

14. There is also mention of an octahedral die being used.

15. Galladay, "*Los libros de acedrex dados e tablas*."

16. David Eugene Smith, *History of Mathematics*, vol. 2 (New York: Dover, 1958), 524.

17. Anicius Manlius Severinus Boethius (480–524 or 525).

18. Nachum Rabinovitch, *Probability and Statistical Inference in Ancient and Medieval Jewish Literature* (Toronto: Toronto University Press, 1973), 144.

19. Smith, *History of Mathematics*, 525.

20. Dorathy M. Robathan, "Introduction to the Pseudo-Ovidian De Vetula," *Transactions and Proceedings of the American Philological Association* 88 (1957): 197–207. See also D. R. Bellhouse, "De Vetula: A Medieval Manuscript Containing Probability Calculations," *International Statistical Review* 68:2 (2000): 123–36.

21. From Thomas Naogeorgus's sixteenth-century poem, *The Popish Kingdome, or, Reigne of Antichrist*, reprinted in Ashton, *A History of Gambling in England*, 50.

22. Stephen Greenblatt, *Will in the World: How Shakespeare Became Shakespeare* (New York: Norton, 2004), 166.

23. Delmar E. Solem, "Some Elizabethan Game Scenes," *Educational Theatre Journal* 6:1 (1954): 21.

24. Aristotle, *De caelo* ii, 6, 8, 10.

Chapter 2
The Professionals

1. These papers remained unpublished for almost a hundred years. See Øystein Ore, *Cardano: The Gambling Scholar* (Princeton: Princeton University Press, 1953. It should be pointed out that this book of Ore's was the first to expose Cardano's contributions to mathematical probability theory.

2. Persi Diaconis, Susan Holmes, and Richard Montgomery, "Dynamical Bias in the Coin Toss," *SIAM Review* 49:2 (2007): 211–35.

3. The law of large numbers is deducible from axioms of probability. Alas, its proof is extremely long and beyond the scope of this book. See W. Feller, *An Introduction to Probability Theory and Its Applications*, vol. 1, 3rd ed. (New York: Wiley, 1971), 243–48.

4. It's not likely that Galileo knew of Cardano's *Liber de Ludo Aleae*.

5. G. Galileo (ca. 1620), *Sopra la scoperte die dadi* (On a discovery concerning dice), trans. E. H. Thorne, excerpted in F. N. David, *Games, Gods, and Gambling: The Origins and History of Probability and Statistical Ideas from the Earliest Times to the Newtonian Era* (New York: Hafner, 1962), 192–95.

6. This was first published in 1663.

7. For more detail and insight on the problem of points, read Keith Devlin's book *The Unfinished Game* (New York: Basic Books, 2008).

8. Øystein Ore, "Pascal and the Invention of Probability Theory," *American Mathematical Monthly* 67:5 (1960): 409–19.

9. The original letters were edited and published in *Oeuvres de Fermat*, ed. Paul Tannery and Charles Henry, vol. 2 (Paris, 1894), 288–314. For the letters

in translation, see David Eugene Smith, *A Source Book in Mathematics* (New York: Dover, 1959), 424.

10. In fact, in the book, Bernoulli acknowledges that much of the first part is a reproduction of Huygens's *De Rationciniis in Ludo Aleae*, and that Leibniz and Wallis already knew the material on permutations and combinations.

11. Oliver Goldsmith, *The Collected Works of Oliver Goldsmith*, vol. 3, ed. Arthur Friedman (Oxford: Oxford University Press, 1966), 311.

12. Ashton, *A History of Gambling in England*, 66–67.

13. *The Guardian*, July 29, 1713.

14. Charles Cotton, *The Compleat Gamester: Or Instructions How to Play at All Manner of Usual and Most Genteel Games* (1674; rept., Barre, MA: Imprint Society, 1970), 14.

CHAPTER 3
FROM COFFEEHOUSES TO CASINOS

1. Russell T. Barnhart, "Gambling in Revolutionary Paris—The Palais Royal: 1789–1838," *Journal of Gambling Studies* 8:2 (1992): 151.

2. Ibid.

3. For a lucid modern history of the transition from tavern to coffeehouse in the time of Oliver Cromwell, see Tom Standage, *A History of the World in 6 Glasses* (New York: Walker, 2005), 141.

4. Frederick Martin, *The History of Lloyd's and of Marine Insurance in Great Britain* (London: Macmillan, 1876), 55.

5. Ralph Nevill, *London Clubs: Their History and Treasures* (London: Chatto and Windus, 1911), 3.

6. White's is still operating at 37 St. James Street.

7. William E. H. Lecky, *History of England in the Eighteenth Century* (New York: D. Appleton, 1890), 556.

8. Alexander Pope, *The Dunciad*, book 1 (Whitefish, MT: Kessinger Publishing, 2004), 9.

9. David Schwartz, "A New Deal: Bruges Burgers and Venetian Merchants Invent Mercantile Gambling," XIV International Economic History Congress, Helsinki, 2006, Session 24, available at http://www.helsinki.fi/iehc2006/papers1/Schwartz.pdf.

10. Barnhart, "Gambling in Revolutionary Paris."

11. Lawrence Stone and Jeanne Stone, *An Open Elite?: England, 1540–1880* (New York: Oxford University Press, 1986), 218.

12. G. Trevelyan, *The Early History of Charles James Fox* (Whitefish, MT: Kessinger Publishing, 1881), 88–89.

13. Horace Walpole, *Letters of Horace Walpole, Earl of Orford to Sir Horace Mann, British Envoy at the Court of Tuscany*, ed. Lord Dover, vol. 2 (New York: George Dearborn, 1833), 233.

14. Though in one scene he is just outside White's, presumably on his way to the club. However, this gambling scene is not at White's but at another club.

15. Steinmetz, *The Gaming Table*, 1:vii.

16. Percy Colson, *White's, 1693–1950* (Oxford: Heinemann, 1951), 100.

17. Steinmetz, *The Gaming Table*, 1:158.

18. Ibid.

19. *New York Times*, Monday, September 14, 1858.

20. David G. Schwartz, *Roll the Bones: The History of Gambling* (New York: Gotham, 2006), 195–96.

21. Originally quoted in Russell T. Barnhart, *Gamblers of Yesteryear* (Las Vegas: GBC Press, 1983), 108; referenced in Schwartz, *Roll the Bones*, 195–96.

22. Steinmetz, *The Gaming Table*, 1:160.

23. For a more complete and lively history of the birth and development of the Monte Carlo casino, see David Schwartz's magnificent book, *Roll the Bones*.

CHAPTER 4
THERE'S NO STOPPING IT NOW

1. Andrew Steinmetz, *The Gaming Table: Its Votaries and Victims, In All Times and Countries, Especially in England and in France*, vol. 1 (London: Tinsley Brothers, 1870), 92.

2. Lisa Burgess, "Buyers Beware: The Real Iraq 'Most Wanted' Cards Are Still Awaiting Distribution," *Stars and Stripes*, European edition, April 17, 2003.

3. This comes from Steinmetz's account of passing 102 The Bowery and watching the young hawker (in *The Gaming Table*, 1:87).

4. The history of the lottery is vague and doubtful. My research found the dates and evidence for lottery history contradictory; however, one secondary source seems to be more believable than others: see Schwartz, *Roll the Bones*, 84–91.

5. C. L'Estrange Ewen, *Lotteries and Sweepstakes: An Historical, Legal, and Ethical Survey of Their Introduction, Suppression, and Re-Establishment in the British Isles* (New York: Benjamin Bloom, 1972), 24–28.

6. Schwartz, *Roll the Bones*, 85.

7. John Samuel Ezell, *Fortune's Merry Wheel: The Lottery in America* (Cambridge, MA: Harvard University Press, 1960), 71, table 4.

8. Ibid., 12.

9. For a complete list, see ibid., 55–59, table 1.

10. Benjamin Franklin organized the Philadelphia Lottery in 1746.

11. Herbert Asbury, *Sucker's Progress: An Informal History of Gambling in America from the Colonies to Canfield* (New York: Dodd, Mead, 1938), 74–76.

12. Ibid., 78.

13. "Gambling Is Never Enough," *New York Times*, May 10, 2008. This editorial asserts: "There are already more than 12,000 video lottery terminals in eight facilities across New York. No matter how much the state rakes in from gambling, it never seems to prevent budget deficits."

14. In the British House of Commons Record, February 11, 1780.

15. Suzanne McGee, "Trading Places: Stock Markets May Look Nothing Like They Used To, But They Still Serve the Same Crucial Role," *Wall Street Journal*, January 11, 1999.

16. Schwartz, *Roll the Bones*, 119.

17. Ashton, *A History of Gambling in England*, 277–81.

18. Carol Loomis, "The Jones Nobody Keeps Up With," *Fortune*, April 1966, 237–47.

19. Steve Johnson, "A Short History of Bankruptcy, Death, Suicides and Fortunes," *Financial Times*, April 27, 2007.

20. By my count and from my interpretations from his biography, Livermore made and lost fortunes five times during his lifetime.

21. *New York Tribune*, November 30, 1940.

22. Jones founded one of the first hedge funds with $100,000. See *New York Times* obituaries, "Alfred W. Jones, 88, Sociologist and Investment Fund Innovator," Friday, May 2, 2008.

23. Frank J. Travers, *Investment Manager Analysis: A Comprehensive Guide to Portfolio Selection, Monitoring, and Optimization* (Hoboken, NJ: John Wiley, 2004), 333.

24. Ezell, *Fortune's Merry Wheel*, 10.

25. David Jenkins, "The Long and Short," *The Guardian*, September 24, 2005.

26. Judy Woodruff, "The Financial Crisis: An Interview with George Soros," *New York Review of Books* 55:8 (May 15, 2008): 8.

CHAPTER 5
BETTING WITH TRILLIONS

1. B. F. Skinner's reason for the increased likelihood that the gambler will continue to play. See chapter 3.

2. James B. Stewart, "The Omen," *The New Yorker*, October 20, 2008, p. 58.

3. Ibid., 63.

4. It was Andy Bechtolsheim, the cofounder of Sun Microsystems. Just six months later, the venture capital firms of Kleiner Perkins Caufield & Byers and

Sequoia Capital brought in substantial capital. See Verne Kopytoff and Dan Fost, "For Early Googlers, Key Word Is $$$," *San Francisco Chronicle*, April 29, 2004.

5. Warren Buffett, "Buy American," *New York Times*, October 17, 2008.

6. Russell Baker, "A Fateful Election," *New York Review of Books*, November 6, 2008, p. 4.

7. Luigi Zingales, "Causes and Effects of the Lehman Brothers Bankruptcy," Testimony before the Committee on Oversight and Government Reform, United States House of Representatives, October 6, 2008.

8. Kathryn Dominguez, Ray Fair, and Matthew Shapiro, "Forecasting the Depression: Harvard versus Yale," *American Economic Review* 78:4 (September 1988): 595–612.

9. See Peter Boone, Simon Johnson, and James Kwak, "The Baseline Scenario," October 20, 2008, http://baselinescenario.com.

10. James Bandler, Roddy Boyd, and Doris Burke, "Hank's Last Stand," *Fortune*, October 7, 2008.

11. Ibid.

12. Mid-Year Market Survey report issued by ISDA (International Swaps & Derivatives Association, Inc.), September 24, 2008, http://www.isda.org/press/press092508.html.

13. Nicholas Varchaver and Katie Benner, "The $55 Trillion Question," *Fortune*, September 30, 2008.

14. Nicholas T. Chan, Mila Getmansky, Shane M. Haas, and Andrew W. Lo, "Systemic Risk and Hedge Funds," MIT Sloan School of Management Research Paper 4535-05, February 2005.

15. See the following for an interesting alternative to the old financial market theory: George Soros, "The Crisis & What to Do about It," *New York Review of Books* 55:19 (December 4, 2008).

16. Andrew Lo, *Hedge Funds: An Analytic Perspective* (Princeton: Princeton University Press, 2008), 199.

17. Henry M. Paulson Jr., "Fighting the Financial Crisis, One Challenge at a Time," *New York Times*, November 17, 2008.

CHAPTER 6
WHO'S GOT A ROYAL FLUSH?

1. For the complete story, see chapters 12 and 13 of my book *Euclid in the Rainforest* (New York: Plume, 2007).

2. R. Fisher, "Mathematics of a Lady Testing Tea," in J. Newman, *The World of Mathematics*, vol. 3 (New York: Simon and Schuster, 1956), 1512–21.

CHAPTER 7
THE BEHAVIOR OF A COIN

1. Du Camp was a founder of the *Revue de Paris*; however, the current *Paris Review* has no historic connection to the *Revue de Paris*, which dissolved in 1853.

2. Gustave Flaubert, *Madame Bovary*, trans. Francis Steegmuller (New York: Random House, 1957), 3–4.

3. This refers to licensed practitioners who were permitted to practice medicine in rural country towns that had a shortage of MDs.

4. A. Tversky and D. Kahneman, "Judgment under Uncertainty: Heuristics and Biases," *Science* 185 (September 1974): 1124.

5. D. Kahneman and A. Tversky, "On the Psychology of Prediction," *Psychological Review* 80 (1973): 237.

6. Tversky and Kahneman, "Judgment under Uncertainty," 1125.

7. Ibid. Numbers in parentheses represent the number of students who chose each answer.

8. Tversky and Kahneman, "Judgment under Uncertainty."

9. Kahneman and Tversky, "On the Psychology of Prediction."

10. This comes from a magnificent interactive demonstration site sponsored by Wolfram Mathematica. See http://demonstrations.wolfram.com/LeadsInCoinTossing/.

11. See Feller, *An Introduction to Probability Theory and Its Applications*, 1:243–48.

12. For a terrific interactive applet demonstrating the law of large numbers, see Professor Philip Stark's Web site at http://www.stat.berkeley.edu/~stark/index.html.

13. The paradox was invented by Nicolas Bernoulli, who first stated it in a letter to Pierre Rémond de Montmort dated September 9, 1713.

14. For a translation of the original Latin essay, see Daniel Bernoulli, "Exposition of a New Theory on the Measurement of Risk," trans. Louise Sommer, *Econometrica* 22:1 (1954): 23–36.

15. This idea is developed in the economic theory of expected utility put forth by John Von Neumann and Oscar Morgenstern in *Theory of Games and Economic Behavior* (Princeton: Princeton University Press, 1980).

16. *Casablanca*, directed by Michael Curtiz (Warner Bros., 1942).

17. Karl Pearson, *The Chances of Death and Other Studies in Evolution* (London: Edward Arnold, 1897), 45.

18. We are talking about roulette in Monaco. American roulette differs from European by including a double-zero (00) slot as well as zero (0). However, the coin-flipping analogy is very similar—the double zero counts as both red and black.

19. Pearson, *The Chances of Death and Other Studies in Evolution*, 55.

20. Ibid., 62.

21. Graham Sharpe, *Gambling's Strangest Moments: Extraordinary But True Stories from over 450 Years of Gambling History* (London: Robson, 2005), 237.

22. After taxes and a $5,000 tip to the croupier.

CHAPTER 8
SOMEONE HAS TO WIN

1. However, to fit it on a page it must be horizontally scaled down to look like the graph in figure 12.3.

2. Note that for roulette $p = 9/19$ and $q = 10/19$. Not a big difference between p and q. But in the case of craps, the probability of getting 7 on any one throw of two dice is 1/6 because there are 6 distinct ways two dice can add up to 7 and there are 36 possible outcomes if we do not distinguish which die has which number. So, $p = 1/6$ and $q = 5/6$.

3. Pearson, *The Chances of Death and Other Studies in Evolution*, 55.

4. Warren Weaver, *Lady Luck: The Theory of Probability* (Garden City, NY: Doubleday, 1963), 282.

5. Abraham De Moivre, *The Doctrine of Chances*, 3rd ed. (1756; repr., New York: Chelsea 1967), 243.

CHAPTER 9
A TRULY ASTONISHING RESULT

1. Jacob Bernoulli, *The Art of Conjecturing*, trans. Edith Dudley Sylla (1713; rept., Baltimore: Johns Hopkins University Press, 2006), 339.

2. For a proof, see Weaver, *Lady Luck*, 232–33.

3. Bernoulli, *The Art of Conjecturing*, 101.

4. This quote appears on page 132 of Edith Dudley Sylla's translation of Bernoulli's *Ars Conjectandi*. Huygens's *De Ratiociniis in Ludo Aleae* is reproduced as part 1 of *Ars Conjectandi*. It actually appeared first as an appendix to a book of mathematical exercises by Frans van Schooten printed in 1657. Huygens's book is not to be confused with Girolamo Cardano's mathematics gambling manual, *Liber de Ludo Aleae*.

5. Cardano's *Liber de Ludo Aleae* was written in the 1500s and published in 1663, whereas Huygens's *De Ratiociniis in Ludo Aleae* was published in 1657. However, the medieval poem *De Vetula*, ascribed to Richard de Fournival, talked about in chapter 1 of this book, contained a short description of which combinations can come from the tossing of three dice without reference to any hint of expected value.

6. Bernoulli, *The Art of Conjecturing*, 132.

7. Diaconis, Holmes, and Montgomery, "Dynamical Bias in the Coin Toss," 211–35.

8. Stephen M. Stigler, *The History of Statistics: The Measurement of Uncertainty before 1900* (Cambridge, MA: Harvard University Press, 1986), 64–65.

9. Ibid., 77.

10. Ibid., 78.

11. Robert Oerter, *The Theory of Almost Everything* (New York: Pi Press, 2006), 82.

12. First he approximated the sum of the coefficients (which were really probabilities) by a definite integral of an exponential function and then expanded the exponential as a power series so he could compute the integral term by term.

13. Bernoulli, *The Art of Conjecturing*, 329.

14. Oerter, *The Theory of Almost Everything*, 84.

Chapter 10
The Skill/Luck Spectrum

The epigraph comes from Cardano's *Liber de Ludo Aleae (The Book on Games of Chance)*, trans. Sydney Henry Gould, as it appears in Øystein Ore's *Cardano*, 189.

1. Excluding many games that are closely related, such as backgammon, baccarat, and faro.

2. Henri Poincaré, *Calcul des probabilités* (Paris: George Carré, 1896), 122–30.

3. See Nick Leeson, *Rogue Trader* (New York: Time Warner, 1997).

4. In July 2009, the Tri-State Megabucks increased the price of a ticket to $2. Of course this doubles the purse.

5. See chapter 6.

6. This rate varies from state to state but is generally between 0.15 and 0.20.

7. James Surowiecki, *The Wisdom of Crowds* (New York: Random House, 2005).

8. Jack Treynor, "Market Efficiency and the Bean Jar Experiment," *Financial Analysts Journal* 43 (1987): 50–53.

9. Arthur E. Hoerl and Hervert K. Fallin, "Reliability of Subjective Evaluations in a High Incentive Situation," *Journal of the Royal Statistical Society* 137 (1974): 227–30.

10. R. M. Griffith, "Odds Adjustments by American Horse-Race Bettors," *American Journal of Psychology* 62:2 (1949): 290–94.

11. John Scarne, *Scarne's Complete Guide to Gambling* (New York: Simon and Schuster, 1961), 33.

12. Edward O. Thorp, *Beat the Dealer* (New York: Vintage, 1966), 39.

13. Ibid., 45.

14. Roger R. Baldwin, William E. Cantey, Herbert Maisel, and James P. McDermott, "The Optimum Strategy in Blackjack," *Journal of the American Statistical Association* 51:275 (1956): 429–39.

15. For another strategy that precedes the "The Optimum Strategy in Blackjack," see E. Culbertson, A. H. Morehead, and G. Mott-Smith, *Culbertson's Card Games Complete* (New York: Greystone Press, 1952).

16. Thorp, *Beat the Dealer*, 18.

17. "A Game of 'Craps' in the Sunny South," *New York Times*, May 14, 1905.

18. Bill Coleman, "How Slots Work: An Insider's Guide to the Inner Workings of Our Favorite—and Most Frustrating—Games," *Strictly Slots*, http://www.strictlyslots.com/archive/0509ss/how_slots_work.html.

CHAPTER 11
LET IT RIDE

1. Dialogue from the movie *High Roller: The Stu Ungar Story* (directed by A. W. Vidmer [New Line Home Video, 2005]), based on a true story about a card prodigy three-time poker champion of the Las Vegas Poker World Series.

2. Fyodor Dostoyevsky, *The Gambler*, trans. C. J. Hogarth (Charleston, SC: Bibliobazaar, 2007).

3. Roulettenberg is modeled on the casino at Wiesbaden, Germany, where Dostoyevsky spent some time when he was writing *The Gambler*.

4. H. Rachlin, "Why Do People Gamble and Keep Gambling Despite Heavy Losses?" *Psychological Science* 1:5 (1990): 294–97.

5. *Pathological Gambling: A Critical Review*, a report of the National Research Council (Washington, DC: National Academy Press, 1999), 241.

6. Ibid., 242.

CHAPTER 12
KNOWING WHEN TO QUIT

The epigraph comes from G. Keren and W. A. Wagenaar, "On the Psychology of Playing Blackjack: Normative and Descriptive Considerations with Implications for Decision Theory," *Journal of Experimental Psychology: General* 114:2 (1985): 133–58.

1. George Eliot, *Daniel Deronda* (New York: John W. Lovell, n.d.), 8–9.

2. Ibid., 5.

3. Dostoyevsky, *The Gambler*, 87.

4. Guido Baltussen, Thierry Post, and Martijn J. van den Assem, "Risky Choice and the Relative Size of States," 2008, http://ssrn.com/abstract=989242.

5. Richard H. Thaler and Eric Johnson, "Gambling with the House Money and Trying to Break Even: The Effects of Prior Outcomes in Risky Choice," *Management Science* 36:6 (1990): 643–44.

6. Mrs. McKee appeared on the episode of *Deal or No Deal* that aired on January 3, 2008.

7. *Deal or No Deal*, episode #335.

8. An equivalent experiment appeared in Thaler and Johnson, "Gambling with the House Money and Trying to Break Even."

9. Economists have many definitions of risk that depend on the application, but here it's being very loosely defined as the probability that the actual return will be different than the expected value. Some textbooks define it as the standard deviation of the historical average of returns.

10. Richard Thaler, "Mental Accounting and Consumer Choice," *Marketing Science* 4:3 (1985): 199–214.

11. Tversky and Kahneman, "Judgment under Uncertainty," 1124–31.

12. Maurice Allais, "Le comportement de l'homme rationnel devant le risque: Critique des postulates et axiomes de L'Ecole Americaine," *Econometrica* 21 (1953): 503–46. See also Maurice Allais and O. Hagen, *Expected Utility Hypothesis and the Allais Paradox* (Dordrecht-Boston: Reidel, 1979).

13. This is a thought experiment based on hypothetical large-size winnings. Many variations of this experiment have been conducted over the years with surprisingly similar results. See http://www.zoology.ubc.ca/~hauert/research/gamelab/allais.html.

14. See Mitchell Zuckoff, "The Perfect Mark," *The New Yorker*, May 15, 2006, pp. 38–42.

15. Ibid., 41.

16. Ibid.

17. Ibid., 42.

CHAPTER 13
THE THEORIES

1. Sigmund Freud, *The Ego and the Id*, reprinted in *The Freud Reader*, ed. Peter Gay (New York: Norton, 1989), 643.

2. Bergler was also known for his contentious belief that homosexuality was a choice and that the "disease" could be cured. Interesting that such a prolific analyst with such an admired school could be altogether wrong about something so central to his field.

3. Dostoyevsky, *The Gambler*, 81.

4. Ibid., 82.

5. Ibid., 36.

6. Theodor Reik, *From Thirty Years with Freud*, trans. Richard Winston (New York: Farrar and Rinehart, 1940), 171.

7. Edmund Bergler, *The Psychology of Gambling* (New York: International Universities Press, 1957), 23.

8. Ibid., 128, and chapter 10.

9. Ibid., 1.

10. Richard Jessup, *The Cincinnati Kid* (New York: Plume, 1985).

11. Dostoyevsky, *The Gambler*, 77.

12. Ibid., 96.

13. Joseph Frank, *Dostoyevsky*, vol. 4: *The Miraculous Years, 1865–1871* (Princeton: Princeton University Press, 1995), 177.

14. See Robert L. Jackson, *Dostoyevsky: New Perspectives* (Englewood Cliffs, NJ: Prentice Hall, 1984), 116.

15. According to Joseph Frank, this might have eaten into his only source of income for a considerable time.

16. According to his wife, Anna Dostoyevsky, "The money was paid out to the creditors the very next day, and so Fyodor Mikhailovich never even held a penny of it in his hands. But the most galling thing of all was that all this money came back to Stellovsky again within a few days. It turned out that he himself had bought up Fyodor Mikhailovich's promissory notes for next to nothing and extracted the money from him through two intermediaries who were really figureheads." Anna Dostoevsky, *Dostoevsky Reminiscences*, trans. Beatrice Stillman (New York: Liveright, 1975).

17. Dostoevsky, *Dostoevsky Reminiscences*, 126–27.

18. See R. R. Greenson, "On Gambling," *American Imago* 4 (1947): 61–77; and J. Gladston, "The Gambler and His Love," *American Journal of Psychiatry* 117 (1960): 553–55.

19. B. F. Skinner, *Science and Human Behavior* (New York: The Free Press, 1953), 397.

20. I'm using the word *reinforcement* here the way Pavlov defined it: a *reinforcement* is any event that strengthens (reinforces) behavior.

21. Skinner, *Science and Human Behavior*.

22. Here, I'm using the word *conditioning* the way Pavlov defined it: *conditioning* is the strengthening of behavior that comes by way of reinforcement.

23. *Pathological Gambling: A Critical Review*, 242.

24. There would be the chance of many more people splitting the winnings—for each winning number there may be several winners, so the likelihood of having many more than a hundred winners is significantly increased.

25. Dostoyevsky, *The Gambler*, 82.

26. R. A. McCormick and J. I. Taber, "The Pathological Gambler: Salient Personality Variables," in T. Galski, ed., *The Handbook of Pathological Gambling* (Springfield, IL: Charles C. Thomas, 1987), 352–53.

27. H. J. Shaffer, D. A. LaPlante, R. A. LaBrie, R. C. Kidman, A. Donato, and M. V. Stanton, "Toward a Syndrome Model of Addiction: Multiple Manifestations, Common Etiology," *Harvard Review of Psychiatry* 12:6 (2004): 367–74.

28. Compare http://www.gamblersanonymous.org/recovery.html and http://www.aa.org/.

29. Joseph Shapiro, "America's Gambling Fever," *U.S. News and World Report*, January 15, 1996.

30. *Pathological Gambling: A Critical Review*, 24.

31. Glen Walters, "The Gambling Lifestyle: I. Theory," *Journal of Gambling Studies* 10:2 (1994): 168.

32. There is a rumor that Senator John McCain keeps a lucky coin and rabbit's foot in his pocket at all times.

33. Even in Chaturanga, the oldest form of chess (dating back to seventh-century India), dice were thrown to determine who had the next move.

34. See Mark Dickerson, John Henchy, Martin Schaefer, Neville Whitworth, and John Fabre, "The Use of a Hand-Held Microcomputer in the Collection of Physiological, Subjective and Behavioral Data in Ecologically Valid Gambling Settings," *Journal of Gambling Behavior* 4:2 (1988): 92–98.

35. Bergler, *The Psychology of Gambling*, 8.

36. "Search for Jeans Brought Jackpot," *New York Daily News*, Saturday, January 26, 2008. See also "Battle of the Jeans Ends in a Jackpot," *Seattle Times*, Saturday, January 26, 2008.

37. See David Oldman, "Chance and Skill: A Study of Roulette," *Sociology* 8:3 (1974): 407–26. Oldman was a roulette croupier before designing field studies on the cognitive behavior of gamblers.

38. Michael Walker, "Irrational Thinking among Slot Machine Players," *Journal of Gambling Studies* 8:3 (1992): 250.

39. We have not talked about Internet gambling in this book. One reason is that at the time of publication, the numbers were insignificant. According to a Gallup survey in 2004, online gambling accounts for less than 1 percent of American gambling. What we do know is that it is a relatively new phenomenon that is growing at a fast pace and there may be some worries about future effects. For example, according to a 2005 British study by Inside Edge, online gambling showed a 566 percent increase in just one year.

40. Michael D. Lemonick, "How We Get Addicted," *Time*, July 5, 2007.

41. Ibid.

42. Terence O'Brian, "Elders and Addiction: Research Paths to Integrated Care," *Aging Today* 28:6 (2007): 7–8.

CHAPTER 14
HOT HANDS

1. I thank my good friend the jazz musician Bobby Previte for telling me this during a chat on the thrills of gambling.

2. P.-S. de Laplace, *Essai philosophique sur les probabilités* (1796), reprinted as *A Philosophical Essay on Probabilities* (New York: Dover, 1951).

3. T. Gilovich, R. Vallone, and A. Tversky, "The Hot Hand in Basketball: On the Misperception of Random Sequences," *Cognitive Psychology* 17 (1985): 295–314.

4. "'Hot Hands' Phenomenon: A Myth?" *New York Times*, April 19, 1988, sec. C, p. 1.

5. Ibid.

6. Dostoyevsky, *The Gambler*, 36.

7. In other words, people make irrational decisions and are not as calculating as standard economic models assume.

8. Kahneman and Tversky, "Judgment under Uncertainty," 1125.

9. D. Kahneman and A. Tversky, "Subjective Probability: A Judgment of Representativeness," *Cognitive Psychology* 3 (1972): 430–54.

10. Willem Wagenaar, "Generation of Random Sequences by Human Subjects: A Critical Survey of Literature," *Psychological Bulletin* 77 (1972): 65–72.

11. Peter Ayton and Ilan Fischer, "The Hot Hand Fallacy and the Gambler's Fallacy: Two Faces of Subjective Randomness?" *Memory and Cognition* 32:8 (2004): 1370.

12. E. Langer, "The Illusion of Control," *Journal of Personality and Social Psychology* 32 (1975): 311–28.

13. James Sundali and Rachel Croson, "Biases in Casino Betting: The Hot Hand and the Gambler's Fallacy," *Judgement and Decision Making* 1:1 (2006): 3.

14. A. Leopard, "Risk Preference in Consecutive Gambling," *Journal of Experimental Psychology: Human Perception and Performance* 4 (1978): 521–28.

15. D. Oldman, "Chance and Skill: A Study of Roulette," *Sociology* 8 (1974): 418. Note: Along with Sundali and Croson (note 13), this essay is one of the rare observational field studies on betting behaviors for roulette. See also Willem Wagenaar, *Paradoxes of Gambling Behavior* (London: Psychology Press, 1989), chapter 4. For observational field studies of biased beliefs in blackjack (a game of mixed skill and luck), see Keren and Wagenaar, "On the Psychology of Playing Blackjack," 133–58.

16. M. D. Griffiths, "The Role of Cognitive Bias and Skill in Fruit Machine Gambling," *British Journal of Psychology* 85 (1994): 351–69.

CHAPTER 15
LUCK

1. It ranks third after the Venetian in Macau and the Mohegan Sun in southeastern Connecticut.

2. This is according to Foxwoods' Web site, http://www.foxwoods.com/AboutFoxwoods.

3. U.S. casinos have been successfully fighting city smoking bans. Given recent neurobiological research on addictions that link addictive behaviors to dependent biopsychosocial disorders, casino managers support the theory that excluding smokers excludes a core of gamblers. One Atlantic City councilman, John Shultz, said this: "Gaming is about smoking, drinking, and gambling. It all goes together. It's all sin." From Las Vegas to Atlantic City, casino executives have argued that profits drop by 10 percent under smoking bans.

Further Reading

General Gambling Information, Including How Games Are Played

Scarne, John. *Scarne's Complete Guide to Gambling.* New York: Simon and Schuster, 1961.

This book is out of print but can be found used online. It contains almost everything you would want to know about gambling from an intelligent point of view. Scarne was a distinguished gambling expert, very well respected by both gambling insiders and academics. Much of the material is out of date, but it still gives a feel of what gambling was like before it became legal all over America. With a terrific sense of the mathematics behind each game, Scarne's book prosaically explains the math clearly without much symbol fuss.

Background History

Schwartz, David. *Roll the Bones: The History of Gambling.* New York: Gotham, 2006.

This is a magnificent book, written by a historian at the University of Nevada's Center for Gaming Research. Schwartz is a masterful writer with a thorough knowledge of gambling in Western Europe as well as the United States. The book is as entertaining as it is thorough with plenty of anecdotes and useful facts. A very readable book with a broad scope that extends from ancient history to Internet gambling.

The History of Old Games

Cotton, Charles. *The Compleat Gamester, Or Instructions How to Play at All Manner of Usual and Most Genteel Games.* Barre, MA: Imprint Society, 1970.

This charming little book was first published in 1674. The Imprint Society has done an admirable service by making it available once again. It was the favorite book on games in the seventeenth century. It describes thirty-eight different games from billiards to basset as well as the countenances and spirits of the gamesters themselves.

The Mathematics

Bernoulli, Jacob. *The Art of Conjecturing.* Trans. Edith Dudley Sylla. Baltimore: Johns Hopkins University Press, 2006.

First published in 1713, this book was the first comprehensive book on the theory of probability. Thanks to the expert translating and editing of Edith Dudley Sylla, this book contains readable accounts of the early mathematics of permutations and combinations. It also contains a readable account—annotated by Bernoulli himself—of Huygens's *Treatise on Reckoning in Games of Chance*, which introduces dice odds and average values.

Packel, Edward. *The Mathematics of Games and Gambling.* 2nd ed. Washington, DC: Mathematical Association of America, 2006.

This is an exceptionally easy-to-read introduction to probability, expectation, and game theory that includes brief analyses of many games across the skill/luck spectrum from roulette to poker. It's a small book with no prerequisites beyond high school algebra. I confidently recommend this book to anyone who wishes to read with remarkable ease about the odds and expectations of individual games of chance.

Index

addiction, xii–xv, 29, 179, 218, 264n4; behavioral analysis and, 155–56, 186, 193–201; big business and, 36, 39–40; Brummell and, 41–42; compulsive gamblers and, 155, 184, 187, 201, 213; dopamine production and, 200; emotional highs and, 154, 194–96, 200; family effects from, 194; greed and, 156 (*see also* greed); loss of savings from, 209–10; neurobiological research on, 193–94, 199–201; pathological gamblers and, 185–89, 193–201; problem gamblers and, 186–87, 190, 193; retirees and, 209–12; as self-medication, 194–96; sensation-seeking and, 194–95; suicide and, 194; ventral tegmental area (VTA) and, 199–200; Western culture and, 186
aggies, 155–59, 224
alcohol, 193, 197, 199–201
Alexandrovna, Polina, 184, 188
Alfonso, I, King of León and Castile, 9–12
algebra, 20–23, 123–25, 227, 239
Allais Paradox, 178–79
Ambassadors, The (Holbein), 239
America: *California v. Cabazon Band of Mission Indians* and, 53; Civil War and, 48, 53; dice and, 7; fascination with lotteries in, 50, 131, 193, 214–15; gambling growth in, 46–53; Mississippi riverboats and, 1, 46–47; New Orleans and, 1, 46–48; patholgical gambling and, 193–94; percentage of population visiting casinos, 193–94; prohibition and, 46; stock market crisis of 2008 and, 59–71
American International Group (AIG), 64, 67–68
American Psychiatric Association, 187

American Revolution, 34, 46, 50–52
anchoring, 86–87
angstlust (desire for punishment), 183
Apianus, Petrus, 239–40
Aristotle, 18, 101
Arithmetic Book, The (Apianus), 239–40
Ars Conjectandi (Bernoulli), 33, 118–19, 123, 126–27, 257n4
Ars Magna (Cardano), 20
Ashton, John, 34
astragals, 5–7, 26f, 122–23, 236
atep, 8, 219
Atlantic City, 53–54, 98–100, 164–65, 199
Augustus, 9
Austria, 43
averages, 21–22; law of, 23; weak law of large numbers and, 118–30

baccarat, 131, 219
backgammon, 27, 132; box and, 58–59, 224; captain and, 58–59; cheating and, 59, 131; chouette and, 58–59, 224; crew and, 58–59, 225; description of, 219; historical perspective on, 7–11; partner splits and, 59
banknotes, 39
banks: closures of, 64–65; credit-default swaps and, 57, 64, 66–71; government bailouts and, 62, 64, 66, 214; Leeson and, 132; nationalization of, 64; ninja loans and, 62–63, 225; sub-prime crisis and, 60, 64–67, 70; tight networks of, 62–63
baseball cards, xii, 158–59
basset, 219, 266
Bear Stearns, 64, 214
behavior: acquired, 196; addiction and, 155–56, 186, 193–201 (*see also* addiction); cheating, 28, 34, 52, 58–59,

Also by Joseph Mazur

Euclid in the Rainforest: Discovering Universal Truth in Logic and Math (2006)

The Motion Paradox: The 2,500-Year-Old Puzzle behind All the Mysteries of Time and Space (2007)

EDITED BY JOSEPH MAZUR

Number: The Language of Science (2007)